EPA/600/R-16/365F
Final
January 2017
www.epa.gov/research

Evaluating Urban Resilience to Climate Change: A Multisector Approach

National Center for Environmental Assessment
Office of Research and Development
U.S. Environmental Protection Agency
Washington, DC 20460

DISCLAIMER

This document has been reviewed in accordance with U.S. Environmental Protection Agency policy and approved for publication. Mention of trade names or commercial products does not constitute endorsement or recommendation for use.

CONTENTS

LIST OF TABLES v
LIST OF FIGURES vi
ACRONYMS AND ABBREVIATIONS vii
PREFACE ix
AUTHORS, CONTRIBUTORS, AND REVIEWERS x
EXECUTIVE SUMMARY xi
1. INTRODUCTION 1
   1.1. MOTIVATION FOR A CLIMATE CHANGE AND URBAN RESILIENCE
       ASSESSMENT FRAMEWORK 1
   1.2. THE DOMESTIC AND INTERNATIONAL POLICY CONTEXT 5
   1.3. OVERVIEW OF THE EPA CONCEPTUAL FRAMEWORK 5
2. BACKGROUND TO CONCEPTUAL FRAMEWORK AND TOOL
   DEVELOPMENT 9
   2.1. MULTICRITERIA ASSESSMENT AND MIXED METHODS 10
   2.2. QUALITATIVE AND QUANTITATIVE INDICATOR DEVELOPMENT 12
       2.2.1. Qualitative Indicator Development 12
       2.2.2. Quantitative Indicator Selection 17
   2.3. EXAMPLES OF THRESHOLDS FROM PEER-REVIEWED LITERATURE 18
   2.4. EXAMPLES OF THRESHOLDS FROM GOVERNMENT
       ORGANIZATIONS 19
   2.5. EXAMPLES OF USING QUARTILES TO ASSIGN THRESHOLDS 20
   2.6. DATA COLLECTION APPROACH 21
3. DISCUSSION AND CONCLUSIONS 22
   3.1. VISUALIZING RESILIENCE 22
   3.2. THE UTILITY OF QUALITATIVE ANALYSIS 24
   3.3. INDICATORS OF RESILIENCE AND THEIR THRESHOLDS 25
   3.4. SPECIFIC CHALLENGES IDENTIFIED THROUGH TOOL APPLICATION 26
       3.4.1. Discussions with Experts and Gathering City-Specific Knowledge 26
       3.4.2. Lack of Data and Spatial/Temporal Data Variability 27
       3.4.3. Sector Interconnectivity 30
       3.4.4. Revisions to Qualitative and Quantitative Indicators 30
       3.4.5. Threshold-Setting 31
       3.4.6. Integrating Qualitative Information 31
   3.5. FUTURE STEPS 32
APPENDIX A. TECHNICAL STEERING COMMITTEE MEMBERS 34
APPENDIX B. PARTICIPANTS 38
APPENDIX C. AGENDAS FOR WORKSHOPS IN WASHINGTON, DC 42
APPENDIX D. WASHINGTON, DC CASE STUDY 46
APPENDIX E. WORCESTER, MA CASE STUDY 87
APPENDIX F. COMPARISON OF RESULTS FOR WASHINGTON, DC AND
   WORCESTER, MA 115
APPENDIX G. QUALITATIVE INDICATORS: ORDERED 119
APPENDIX H. QUANTITATIVE INDICATORS: ORDERED 129
APPENDIX I. QUALITATIVE INDICATORS: TEMPLATE 139

APPENDIX J. QUANTITATIVE INDICATORS: TEMPLATE	232
APPENDIX K. QUALITATIVE INDICATORS: WASHINGTON, DC	330
APPENDIX L. QUANTITATIVE INDICATORS: WASHINGTON, DC	419
APPENDIX M. QUALITATIVE INDICATORS: WORCESTER, MA	525
APPENDIX N. QUANTITATIVE INDICATORS: WORCESTER, MA	612
REFERENCES	665

LIST OF TABLES

Table 1. Example frameworks to assess community resilience to climate change 3
Table 2. Potential climate changes of concern for urban areas..13
Table 3. Water sector questions related to drought sensitivity, response, and learning. 15
Table 4. Example qualitative and quantitative indicator from urban resilience tool 17
Table 5. Macroinvertebrate Index of Biotic Condition thresholds 18
Table 6. Palmer Drought Severity Index (PDSI) thresholds...19
Table 7. Physical Habitat Index thresholds...20
Table 8. Mobility management (yearly congestion costs saved by operational treatments per capita) thresholds and scores 20
Table 9. Percent access to transportation stops thresholds and scores 21
Table 10. Worcester, MA data availability..28
Table 11. Quantitative data limitations..29
Table 12. Major weather events and their impacts in the District of Columbia since 2003.........49
Table 13. Major weather and other events and their impacts in Worcester, MA 89
Table 14. Washington, DC and Worcester, MA metrics at a glance 116

LIST OF FIGURES

Figure 1. Urban climate resilience framework. 6
Figure 2. Sample quadrant plot. 23
Figure 3. Washington, DC: Average qualitative indicator resilience and importance. 64
Figure 4. Washington, DC: Average quantitative indicator resilience and importance. 64
Figure 5. Washington, DC: Qualitative indicator quadrant mapping. 66
Figure 6. Washington, DC: Quantitative indicator quadrant mapping. 67
Figure 7. Washington, DC economy sector: Qualitative and quantitative indicator quadrant mapping. 68
Figure 8. Washington, DC, energy sector: Qualitative and quantitative indicator quadrant mapping. 70
Figure 9. Washington, DC land use/land cover sector: Qualitative and quantitative indicator quadrant mapping. 72
Figure 10. Washington, DC natural environment sector: Qualitative and quantitative indicator quadrant mapping. 75
Figure 11. Washington, DC people sector: Qualitative and quantitative indicator quadrant mapping. 77
Figure 12. Washington, DC telecommunications sector: Qualitative and quantitative indicator quadrant mapping. 80
Figure 13. Washington, DC transportation sector: Qualitative and quantitative indicator quadrant mapping. 82
Figure 14. Washington, DC water sector: Qualitative and quantitative indicator quadrant mapping. 85
Figure 15. Worcester, MA: Average qualitative indicator resilience and importance. 94
Figure 16. Worcester, MA: Average quantitative indicator resilience and importance. 95
Figure 17. Worcester, MA: Qualitative indicator quadrant mapping. 96
Figure 18. Worcester, MA: Quantitative indicator quadrant mapping. 97
Figure 19. Worcester, MA economy sector: Qualitative and quantitative indicator quadrant mapping. 98
Figure 20. Worcester, MA energy sector: Qualitative and quantitative indicator quadrant mapping. 100
Figure 21. Worcester, MA land use/land cover sector: Qualitative and quantitative indicator quadrant mapping. 102
Figure 22. Worcester, MA natural environment sector: Qualitative and quantitative indicator quadrant mapping. 104
Figure 23. Worcester, MA people sector: Qualitative and quantitative indicator quadrant mapping. 106
Figure 24. Worcester, MA telecommunications sector: Qualitative and quantitative indicator quadrant mapping. 108
Figure 25. Worcester, MA transportation sector: Qualitative and quantitative indicator quadrant mapping. 110
Figure 26. Worcester, MA water sector: Qualitative and quantitative indicator quadrant mapping. 112
Figure 27. Washington, DC and Worcester, MA: Average quantitative indicator and qualitative indicator score quadrant mapping. 118

ACRONYMS AND ABBREVIATIONS

CMRPC	Central Massachusetts Regional Planning Commission
DCWASA	District of Columbia Water and Sewer Authority
DDOE	District Department of Environment
DDOT	District Department of Transportation
EMS	emergency medical service
FEMA	Federal Emergency Management Agency
GEF	Global Environment Facility
GIS	geographic information system
DC HSEMA	District of Columbia Homeland Security and Emergency Management Agency
IPCC	Intergovernmental Panel on Climate Change
LEED	Leadership in Energy and Environmental Design
MCA	multicriteria assessment
MWCOG	Metropolitan Washington Council of Governments
NCPC	National Capital Planning Commission
ORD	Office of Research and Development
PEPCO	Potomac Electric Power Company
PHI	Physical Habitat Index
TSC	Technical Steering Committee
UBPAD	Upper Blackstone Pollution Abatement District
UFA	Urban Forestry Administration
UNFCCC	United Nations Framework Convention on Climate Change
U.S. DOT	U.S. Department of Transportation

USACE	U.S. Army Corps of Engineers
WARN	water/wastewater agency response network
WMATA	Washington Metropolitan Area Transit Authority
WWTP	wastewater treatment plant

PREFACE

This report was prepared by the U.S. Environmental Protection Agency's (EPA's) Air, Climate, and Energy (ACE) research program, located within the Office of Research and Development, with support from the Cadmus Group. The ACE research program provides scientific information and tools to support EPA's strategic goal of taking action on climate change in a sustainable manner. Such action includes both mitigation, which involves reductions in the movement of heat-trapping greenhouse gases into the atmosphere, and adaptation, which involves preparing for and adjusting to the expected future climate. Both are important, but this report focuses on adaptation to climate change. Climate change impacts are diverse, long-term, and not easy to predict. Adapting to climate change is difficult because it requires making context-specific and forward-looking decisions regarding a variety of climate change impacts and vulnerabilities when the future is highly uncertain. Cities are on the front line for responding to potential climate change impacts, but often do not know precisely the qualities or characteristics that make them vulnerable or resilient to different impacts. This report supports the goal of taking action on climate change in a sustainable manner by developing a conceptual framework of urban resilience to climate change and using rigorously selected indicators to assess community resilience to climate change. This framework is then successfully applied to two different communities (Washington, DC and Worcester, MA) to evaluate their levels of resilience to climate change. Results support the usefulness of this indicator-based approach in identifying traits that enhance or inhibit each community's resilience to focus adaptation planning on issues and areas that are least resilient to climate change impacts.

AUTHORS, CONTRIBUTORS, AND REVIEWERS

The Air, Climate and Energy (ACE) research program of EPA's Office of Research and Development was responsible for producing this report. The report was prepared by The Cadmus Group, Inc. in Waltham, MA, under EPA Contract No. EP-C-11-039; Task Order 10. Susan Julius served as the Task Order Project Officer, providing overall direction and technical assistance, and was a contributing author.

prepared by:

Julie Blue, The Cadmus Group, Inc.

Nupur Hiremath, The Cadmus Group, Inc.

Carolyn Gillette, The Cadmus Group, Inc.

Susan Julius, U.S. EPA, Office of Research and Development

INTERNAL REVIEWERS:

Britta Bierwagen, U.S. EPA, Office of Research and Development

Chris Clark, U.S. EPA, Office of Research and Development

Cathy Allen, U.S. EPA, Office of Policy

Kate Johnson, District Department of the Environment

Marissa Liang, Oak Ridge Institute for Science and Education Fellow

EXTERNAL REVIEWERS:

Vincent Lee PE, ARUP

Eric Smith, AICP

Fahim Tonmoy, Griffith University

ACKNOWLEDGMENTS:

We would like to thank the members of the Technical Steering Committee (listed in Appendix A), Washington, DC workshop participants, our collaborators at the District Department of the Environment (DDOE), and those who spoke to us on behalf of the city of Worcester, MA for their assistance during this project.

EXECUTIVE SUMMARY

Global greenhouse gas emissions continue to rise and have been shown to lead to a range of major and potentially adverse effects on the environment and public welfare. One of the objectives of the Office of Research and Development (ORD) within the U.S. Environmental Protection Agency (EPA) is to provide the scientific basis for climate adaptation choices that support sustainable, resilient solutions at individual, community, regional, and national scales. To support this objective, ORD developed a tool that measures urban communities' resilience to climate change. The tool incorporates both indicator data and input from local sector managers to assess urban resilience for eight municipal management sectors: (1) water, (2) energy, (3) transportation, (4) people (public health and emergency response), (5) economy, (6) land use/land cover, (7) the natural environment, and (8) telecommunications. The tool is intended to provide local-level managers with a way to prioritize threats to resilience using locally available data across multiple sectors to inform adaptation planning. This report describes the tool in detail and discusses the results of applying it in two communities as case study examples: Washington, DC and Worcester, MA. The applications are intended to help individual communities as well as to identify important characteristics and activities that can be transferred across communities to strengthen adaptive capacity at the national scale.

URBAN RESILIENCE DEFINITION, CONCEPTUAL FRAMEWORK, AND TOOL

A conceptual framework was developed based on our definition of urban climate resilience: a city's ability to reduce exposure and sensitivity to, and recover and learn from gradual climatic changes or extreme climate events. This ability comes from a city's risk reduction and response capacity, and includes retaining or improving physical, social, institutional, environmental, and governance structures within a city. The components of urban climate resilience reflected in the conceptual framework include three measures of vulnerability (exposure, sensitivity, and response capacity), as well as the process of initiating responsive action, learning from mistakes or ineffective responses, and building risk reduction capacity (reducing exposure and sensitivity, and increasing response capacity). This cycle is supported or affected by the presence of bridges to action (unforeseen, huge leaps made in response and recovery capabilities), barriers to learning, and barriers to responding. These components guided the selection of urban climate resilience indicators for the tool.

Because data were unavailable for some types of information identified by the conceptual framework, a series of questions for local sector managers were developed to reflect factors affecting resilience for which no indicators or appropriate data sets existed. A Technical Steering Committee (TSC) guided the selection of questions for local sector managers and the selection of quantitative indicators best suited to determine climate resilience for each climatic change/event of concern that a city might have, and for each urban service potentially exposed. Questions were developed as qualitative indicators primarily for assessing the abilities of the appropriate city sectors to respond to climate changes/events and to reduce future exposure/sensitivity, enhance response capacity, and learn from past and future experiences.

The assessment approach the project team chose—using quantitative and qualitative data—makes use of detailed data sets when they are available, but recognizes that important elements of a city's resilience would be neglected if qualitative information provided by city managers were excluded. For both the quantitative and qualitative resilience indicators, participants assigned an importance weight of 1 through 4. A weight of 1 indicates low importance, and a weight of 4 indicates high importance. To score the qualitative indicators, four possible answers were developed for each indicator, with each indicator corresponding to a resilience score of 1 through 4 (with 1 representing low resilience and 4 representing high resilience). To score the quantitative indicators, four quantitative ranges were applied to the data associated with each indicator. These ranges also corresponded with a resilience score of 1 through 4. Participants then selected the scores for the qualitative indicators and reviewed the quantitative indicator ranges and corresponding resilience scores. Qualitative and quantitative indicators with high importance weights and high resilience scores demonstrate where cities are most resilient overall. Qualitative and quantitative indicators with high importance weights and low resilience scores demonstrate where cities are least resilient. Areas of city performance with these combinations of rankings are the most critical areas of focus and warrant attention as soon as possible.

Using published literature, threshold values were established for each indicator that defined the upper and lower boundaries of the four resilience categories. These thresholds were designed to represent resilience levels across all U.S. cities. When threshold values were not available in the literature, panel data for U.S. cities were used. If data for an indicator were not available for a sample of U.S. cities, case studies from one or several cities were analyzed to determine the level of resilience for those cases and the representativeness of that indicator for all U.S. cities.

DISCUSSION AND CONCLUSION

The tool was applied in Worcester, MA and Washington, DC, cities representing different endpoints of a broad spectrum of resources, planning, and risk. The use of these contrasting cities as case studies allows for other cities on this spectrum to understand the applications and potential outcomes of using the tool. It also allows us to test the strengths and weaknesses of the tool methodology in a wide range of conditions and provides preliminary insight into the variety of risk exposures across cities with different geographic, economic, population, and historical characteristics.

This project resulted in a comprehensive, transparent, and flexible tool for identifying the greatest risks, successes, and priorities for decreasing urban vulnerability and increasing resilience to climate change. The results can easily be analyzed with respect to the concepts of exposure/sensitivity, response capacity, or learning, as the qualitative and quantitative indicators are characterized accordingly. The visualizations developed to accompany the results of the application of the tool in Washington, DC and Worcester, MA facilitate the interpretation of case study results and are intended to further assist city managers in moving to the next step of implementing climate change adaptation activities.

The data collected may be analyzed in the context of the framework for the purposes of identifying and prioritizing adaptation activities. This prioritization process may involve categorizing critical vulnerabilities (i.e., sectors and issues within sectors for which resilience is low, but importance is high) into issues that can be addressed in a straightforward manner with adaptation planning and implementation, versus those over which there is less control.

The flexibility of the conceptual framework and tool were enhanced by the fact that use of the tool does not necessarily require quantitative data. Indeed, the project team found that the qualitative indicators were essential to the analysis, even when quantitative data were readily available. The qualitative indicators can be mapped to specific events or types of events, providing city managers and planners with a way to identify feedbacks and learn over time. Additionally, the application of the qualitative indicators fosters and requires interaction with and between sector stakeholders, providing greater learning and coordination opportunities that can be used to further refine the resilience assessments and prioritize activities in response to the assessment findings.

Beyond the numeric values of resilience and importance collected across the sectors evaluated (and the supporting data or responses that contributed to those scores), this effort collected important information regarding the challenges that emerged for the knowledgeable professionals in identifying and confirming appropriate and relevant sources of data to effectively assess the proposed indicators. The disparities in data available between the two cities both complicate the data analysis effort and, for cities lacking data, is a telling indicator of potential vulnerabilities to climate change.

Major challenges encountered while developing and applying the tool included: the need to gather city-specific knowledge (and reasonable subjective knowledge); the lack of data for some sectors and the fact of temporal data variability; the need to adequately identify and capture the interconnectivity of sectors and the specific vulnerabilities that may exist as a result of interconnectivities; the need to assess the adequacy or specificity of qualitative and quantitative indicators; and the need to establish reasonable thresholds for all indicators.

Expansion and refinement of the tool remains to be done. For example, much of the remaining work on the interdependencies among the sectors has not been undertaken by this project. Future advancements in our understanding of these interdependencies can be made by examining linkages more closely, such as those between the water and the energy sectors. However, interdependencies have been addressed to some extent in that some qualitative and quantitative indicators have been assigned to more than one sector, when appropriate.

Ultimately, this urban resilience assessment tool offers valuable insight into the resilience of Washington, DC and Worcester, MA, and assessment results can be meaningfully incorporated into ongoing planning. However, in many cases the information provided by the tool yielded as many new questions as answers. With new patterns of more extreme weather across the globe, adaptation is essential for urban communities and should be guided by an assessment of sector-specific and overall resilience to climate change. Potential future expansions or applications of this tool include: adapting it for online use; conducting additional case studies that focus on new users and expanded geographies and examine the potential for pooling multiple communities' resources in the face of shared risk; and sharing the lessons learned and best practices that emerge from the tool's application in specific communities.

1. INTRODUCTION

1.1. MOTIVATION FOR A CLIMATE CHANGE AND URBAN RESILIENCE ASSESSMENT FRAMEWORK

Resilience has generally been defined as the ability to withstand or to recover from adverse circumstances. This concept has been used in a number of fields, including engineering and environmental sciences (Anderies, 2014; Hopkins, 2010). Many different definitions exist, although common themes that run through those definitions include the degree of disturbance that can be tolerated before function is compromised and the capacity to recover rapidly from a physical disturbance (CARRI, 2013).

More recently in the sociological literature, resilience is treated as the ability of a socioecological system to change and improve in response to stress, rather than merely reverting to a steady state (Simmie and Martin, 2010). This is referred to as evolutionary resilience, where evolution refers to the ability to be flexible, diverse, and employ adaptive learning in the context of changing circumstances. Evolutionary resilience frames recovery as a dynamic path of an interrelated system progressing nonlinearly toward one of potentially multiple equilibria, rather than a direct path toward a single equilibrium (Kim and Lim, 2016).

When resilience entered the lexicon of climate change research, it was defined as the "amount of change a system can undergo without changing state" (IPCC, 2001). Vulnerability was viewed as the inverse of resilience and was defined as "the degree to which a system is susceptible to, or unable to cope with, adverse effects of climate change" (IPCC, 1997). Since 2011, the definition has been evolving. First, the Intergovernmental Panel on Climate Change (IPCC) (2012) broadened the definition to include hazards and to introduce a focus on short-term disruptions as well as long-term changes in averages. In 2014, the IPCC definition was further expanded to include evolution in the ability to adapt, as well as learning and transformation (IPCC, 2014), similar to the sociological definition.

These recent modifications to the definition of resilience has allowed the climate change community to better link the issue of climate change with sustainable development. As far back as 2001, the IPCC recognized that adaptive capacity and sustainable development were linked (IPCC, 2001). Resilience, with its inclusion of future states (i.e., not just bouncing back but bounding forward), provides a more robust linkage and theoretical underpinning. Developing climate-resilient pathways requires sustainable-development trajectories that also include adaptation and mitigation to reduce climate change and its impacts.

This concept of climate resilience is key to preparing cities for the impacts of gradual climate change and associated extreme climate events. Increasing populations within urban ecosystems are putting heavier demands on the supporting biophysical and socioeconomic systems (UN, 2014; UN-Habitat, 2011), and their activities are influencing natural systems, serving as forces for environmental change at local, regional, national, and global scales (IPCC, 2014). Climate change represents yet another source of vulnerability for both our natural and human systems.

For urban ecosystems, the IPCC (2014) identifies one of the greatest threats of climate change to be changes in the intensity and frequency of extreme weather events. Such events can severely damage infrastructure and cause economic losses and injury or death to the population within an urban ecosystem. Those urban areas that are along the coast may experience a combination of sea level rise threatening water supplies and infrastructure damage from intense storms (Crosett et al., 2004). The vulnerability of urban ecosystems is expected to be greater in coastal and riverine areas and in areas whose economies are closely linked with climate-sensitive resources, such as agricultural and forest products. Higher temperatures would affect urban air quality, human health, energy and water requirements, and infrastructure. Urban ecosystems in the Southwest, the Mountain West, the Southeast, and the Great Lakes may experience increased strain on water resources due to pervasive drought conditions (USGCRP, 2014). Jenerette and Larsen (2006) illustrate the susceptibility of many cities to climate change, particularly those in more arid environments in which certain provisioning services, such as fresh water, may not be feasibly obtained in sufficient quantity and at affordable rates. Finally, areas that experience increases in annual precipitation and more intense precipitation events would have increased runoff volume and thus, in addition to increased flooding, greater amounts of nonpoint source contamination in their water bodies.

The nonlinear, complex, and dynamic nature of climate change; urban socioeconomic and environmental systems; and their responses poses significant challenges for existing methods and frameworks. It is yet to be seen whether these frameworks are adequate to meet the challenges (Kim and Lim, 2016). Because of the convergence of population centers and exposures to climatic changes, particularly extreme events, developing approaches to analyze the degree of resilience to these events to support planning efforts is needed and could significantly reduce the risks posed by climate change.

There are a number of nascent efforts to develop robust indicator-based frameworks to measure cities' resilience to the complex and dynamic risks posed by climate change in order to inform adaptation planning (Bahadur, 2015; Schipper and Langston, 2015). Table 1 below, adapted from Schipper and Langston (2015), provides a sample of indicator framework efforts, their scope, and the concepts of resilience they are designed to address. All of these frameworks go beyond merely addressing the climate risk, natural hazards, and physical environment to incorporate socioeconomic, learning, and evolutionary aspects of resilience (Schipper and Langston, 2015). The conceptual framework developed and applied by EPA for this project, discussed in more detail in Section 1.3, shifts the focus from domestic and international development to planning at a city and sector level. Additionally, the evolutionary nature of resilience is acknowledged and reflected within the conceptual framework and corresponding tool.

Table 1. Example frameworks to assess community resilience to climate change

Framework	Scope	Resilience Concept	Indicator Approach
Rockefeller Foundation's 100 Resilient Cities (Arup's City Resilience Framework) http://www.100resilientcities.org/resilience#/-_/ http://publications.arup.com/publications/c/city_resilience_framework	Health and wellbeing, economy and society, infrastructure and environment, leadership and strategy	"The capacity of individuals, communities, institutions, businesses, and systems within a city to survive, adapt, and grow no matter what kinds of chronic stresses and acute shocks they experience."	Qualitative indicators that are "driver" statements representing actions that improve cities' resilience (e.g., Infrastructure: provide reliable communication and mobility; Economy and Society: ensure social stability, security, and justice)
Assessments of Impacts and Adaptation of Climate Change (AIACC) sustainable livelihood approach http://www.start.org/Projects/AIACC_Project/working_papers/Working%20Papers/AIACC_WP_No017.pdf	Natural capital, financial capital, physical capital, human capital, social capital	Improving the quality of life without compromising livelihood options for others	Quantitative and qualitative indicators that measure communities' ability to cope with and recover from shocks and stresses, economic efficiency and income stability, ecological integrity, and social equity
UK Department for International Development Building Resilience and Adaptation to Climate Extremes and Disasters (BRACED) framework http://www.braced.org/resources/i/?id=cd95acf8-68dd-4f48-9b41-24543f69f9f1	Adaptive capacity (assets and income; strength and adaptability of livelihoods, availability and use of climate change information, basic services for vulnerable populations), anticipatory capacity (preparedness and planning, capacity, coordination and mobilization, risk information), absorptive capacity (savings and safety nets, substitutable and diverse assets and resources), transformation (leadership, empowerment, and decision-making processes; strategic planning and policy; innovative processes and technologies)	"Ability to anticipate, avoid, plan for, cope with, recover from and adapt to (climate related) shocks and stresses."	Quantitative and qualitative indicators that span climate change impacts data, economic data, livelihood data, ecological data, social and institutional data, and data on planning and decision making processes
United Nations Development Programme's (UNDP's) Community-Based	Natural capital, financial capital, physical capital, human capital, social capital	"Inherent as well as acquired condition achieved by managing risks over time at individual,	"Composite set of context-specific multisectoral quantitative and qualitative resilience indicators." This process tool enables

Framework	Scope	Resilience Concept	Indicator Approach
Resilience Analysis (CoBRA) framework http://www.undp.org/content/undp/en/home/librarypage/environment-energy/sustainable_land_management/CoBRA/cobra-conceptual-framework.html		household, community and societal levels in ways that minimize costs, build capacity to manage and sustain development momentum, and maximize transformative potential, […] and manage change by maintaining or transforming living standards in the face of shocks or stresses without compromising their long-term prospects."	communities to identify key building blocks of resilience and assess the attribution of various interventions in attaining resilience characteristics.
Characteristics of a Disaster Resilient Community http://community.eldis.org/59e907ee/Characteristics2EDITION.pdf	Five thematic areas: governance, risk assessment, knowledge and education, risk management and vulnerability reduction, and disaster preparedness and response	"The capacity to (1) anticipate, minimize and absorb potential stresses or destructive forces through adaptation or resistance; (2) manage or maintain certain basic functions and structures during disastrous events; and (3) recover or 'bounce back' after an event."	Multiple dimensions for analysis, guided by the five thematic areas and three subdimensions (components of resilience, characteristics of a disaster-resilient community, characteristics of an enabling environment). Specific resilience indicators are at the level of activities, such as hazards/risk data and assessment; public awareness; knowledge and skills; financial instruments; early warning systems; and so forth.
United States Agency for International Development (USAID) Measurement for Community Resilience https://agrilinks.org/sites/default/files/resource/files/FTF%20Learning_Agenda_Community_Resilience_Oct%202013.pdf	Food security, nutrition, health, social capital (bonding social capital, bridging social capital, linking social capital), assets, ecosystem health, poverty	"The general capacity of a community to absorb change, seize opportunity to improve living standards, and to transform livelihood systems while sustaining the natural resource base. It is determined by community capacity for collective action as well as its ability for problem solving and consensus building to negotiate coordinated response."	Combination of outcome measures and process measures to establish a baseline food security/nutrition index, health index, asset index, social capital index, and economic/poverty index. Baseline values are reanalyzed after considering the nature of potential shocks and stresses, community capacities to measure resilience, and areas of collective action (e.g., disaster risk reduction, conflict management, social protection, natural resource management, management of public good and services).

1.2. THE DOMESTIC AND INTERNATIONAL POLICY CONTEXT

Efforts to develop frameworks and tools to assess climate change resilience are supported and driven by legislation and policy actions at the local, state, federal, and international levels. The IPCC and the United Nations Framework Convention on Climate Change (UNFCCC) have identified climate change adaptation planning as a key element of the response to climate change on a global level (IPCC, 2007a; UNFCCC, 2010). Nationally, Executive Order 13693 requires consideration of climate change impacts on operations and major facilities, in addition to national emissions reductions (Exec. Order 13693, 2015). Focusing more on vulnerability and adaptation, Executive Order 13690 (2015) requires capital projects funded with taxpayer dollars to include consideration of increasing flood severity. Currently, nearly 40 federal agencies have produced climate change adaptation plans, vulnerability assessments, or metrics (Leggett, 2015), although many of these are high-level or preliminary efforts.

Below the federal level, the majority of states have some climate planning statute (CES, 2014). For example, New York's Community Risk Reduction and Resiliency Act (S6617B, 2014) requires that all projects receiving state money consider the impacts of climate change during the planning process. In 2012, Hurricane Sandy helped highlight infrastructure vulnerability to natural disasters in New York State, encouraging the passage of S6617B, which requires consideration of sea level rise, storm surge, and flooding in new developments, and infrastructure regulations, permits, and funding. At a local level, city governments must prepare for climate change by protecting natural systems, the built environment, and the human population (Carmin et al., 2012). Sixty-eight percentage of cities worldwide have recognized the importance of preparing for climate change and are in various stages of preparing or implementing adaptation plans (Carmin et al., 2012). Currently in the United States, local governments or agencies in 21 states have developed a total of 66 adaptation plans (Georgetown Climate Center, 2014).

1.3. OVERVIEW OF THE EPA CONCEPTUAL FRAMEWORK

Consistent with the underlying principles in the literature and embodied in the frameworks in Table 1, EPA developed a framework depicting the elements of resilience of an urban system (see Figure 1). The framework builds on our definition of urban resilience to climate change (see Box 1 for a list of working definitions) and employs a hybrid approach that uses both quantitative and qualitative information to assess resilience. The framework not only includes the concepts of vulnerability, exposure, and hazards that present risks to urban environments, but it also goes beyond a static view of the world and incorporates the concepts of feedbacks, learning over time, and evolving in the ability to adapt and respond to challenges presented by gradual and extreme climate change. The framework represents an ongoing process rather than a temporary state of response to external shocks (similar to Engle et al., 2013).

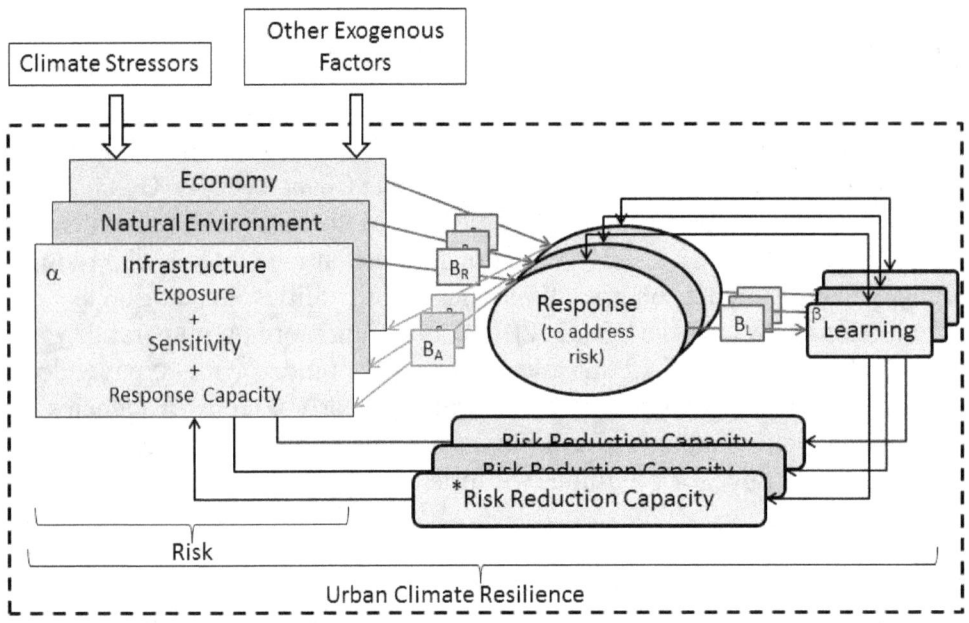

B_A = Bridges to action
B_L = Barriers to learning
B_R = Barriers to responding

Figure 1. Urban climate resilience framework.
α = These three elements—exposure, sensitivity, and response capacity—compose urban vulnerability.
β = Learning outcomes are on three levels: reacting, reframing, and transforming (see Figure 1-3, IPCC, 2012). Examples: reacting = increase a levee height; reframing = realizing the need to assess new storm duration frequency distributions; transforming = assessing societal constructs and migrating to a more robust and comprehensive risk management strategy.
*Risk reduction capacity is the ability to reduce exposure, reduce sensitivity, and/or increase the system's inherent recovery potential in anticipation of harmful climatic changes/events.

In the framework above, the left-hand side focuses on anticipated future climate events and system responses. Included here are the potential exposures to climate change, both gradual and extreme, the potential sensitivity of sectors and systems to those exposures, and the theoretical capability to respond to anticipated climate changes (response capacity, also referred to as adaptive capacity in the climate change literature). The right side of the framework reflects actual responses to real-world experiences of extreme weather events or gradual changes in climate (whether by a community or through observations of other communities and their experiences). Barriers to action and bridges to better-than-anticipated responses are identified based on reflections after an event has occurred. The framework is meant to be applied iteratively through time to capture the forward-looking, dynamic aspect of climate change and planning.

Box 1. Working definitions.

Urban climate resilience: The ability of a city or urban system, through its risk reduction and response capacity, to reduce exposure and sensitivity to, and recover and learn from, gradual climatic changes or extreme climate events, in order to retain or improve the integrity of its infrastructure and economic systems; vital environmental services and resources; the health and welfare of its populations and communities; and the flexibility and diversity of its institutional and governance structures (adapted from Leichenko, 2011).

Exposure: The presence of people; livelihoods; environmental services and resources; infrastructure; or economic, social, or cultural assets in places that could be affected by climate change stressors (adapted from IPCC, 2012).

Sensitivity: Predisposition of human beings, infrastructure, society, and ecosystems to be affected by exposure to a climate stressor or an effect of that exposure (adapted from IPCC, 2012).

Response capacity: Intrinsic capacity of a community to recover from alterations in its normal functioning due to gradual changes in the climate or to extreme events that result in adverse human, material, economic, or environmental effects.

Learning: Ability to recognize complex dynamics of socio-ecological systems in order to respond appropriately to risk and make effective adaptation responses, identify mistakes and shortcomings in those responses following climate stressor events, and evolve as new information becomes available (drawn from IPCC, 2012; Kasperson, 2012).

Bridges to action: Conditions under which unforeseen and huge leaps are made in a community's ability to respond to and recover from alterations or disruptions in its normal functioning (e.g., due to social or technical innovation).

Risk: A function of the exposure to and severity of the occurrence of a particular type of climate change (gradual or extreme) and the way in which its consequences are likely to be mediated by the social vulnerability of the human system. Risk can be assessed in terms of condition and predictive variables representing factors such as economic well-being; health and education status; and preparedness and coping ability with respect to particular climatic changes.

Risk reduction capacity: Ability to reduce risk by reducing exposure and sensitivity or increasing recovery potential and adaptive capacity to prepare for expected climatic changes or events.

Note: These are considered operational definitions. They have been selected for their appropriateness to this application, even though they might not be identical to the definitions in the current literature.

Increasing the resilience of urban environments to climate impacts can happen on both sides of the framework through reducing exposure or sensitivity of systems to potential impacts, expanding the response capacity, increasing learning, removing barriers that inhibit good responses, and providing bridges to promote greater-than-anticipated responses.

This framework serves as the basis for determining the type and breadth of indicators needed to assess a city's resilience condition and evolution over time. Both quantitative and qualitative indicators are employed to capture the various components and processes in the framework. This hybrid approach provides more flexibility in the types and sources of information that can be used, and it reduces bias that can be present if there are limited quantitative data sets available for specific places (Engle et al., 2013). Resilience research and indicators have often been based on quantitative information alone, but interactions with stakeholders (city planners) can lead to important qualitative information that considers local context, refines understanding of specific local vulnerability and resilience, and can calibrate and verify indicators (Engle et al., 2013).

The indicators selected are mapped to specific gradual and extreme climate events facing cities, and to sectors within cities in order to better inform decision makers at the local level. The framework provides thresholds established from the literature against which to measure resilience, rather than relying on measures based on comparisons to other cities (relative resilience). The value of understanding resilience (or lack thereof) is in using that information to take action to avoid or move farther from and above thresholds in order to grow resilience. Our approach provides for flexibility in the final selection of indicators, which allows communities to tailor the resilience assessment to local situations. These innovative features combine elements from other frameworks but do not reside in any other single framework.

The remainder of this report describes in more depth the process of applying this framework to the selection of qualitative and quantitative indicators (see Chapter 2) and developing the tool to assess urban resilience to climate change (see Chapter 2). The report then discusses the results and general insights from applying the tool to two case studies (see Chapter 3). Detailed results of the two case studies, Washington, DC and Worcester, MA, are provided in Appendices D and E, respectively, and a comparison of results across the two case studies is provided in Appendix F.

2. BACKGROUND TO CONCEPTUAL FRAMEWORK AND TOOL DEVELOPMENT

The conceptual framework shown in Figure 1 captures the critical elements of resilience and shows the boundaries (both spatial and conceptual) of our analysis (e.g., the region and beyond are exogenous). This framework guided the selection of urban climate resilience indicators used in our tool and tested at both case study sites. The project team tested and refined the tool in Washington, DC and Worcester, MA to support work in those two communities and to create a guide for city planners and others in other urban environments.

The framework is meant to be applied periodically to capture changes in resilience over time as decision makers enact policies and take action to increase resilience to climate change. The project team first outlined the full array of climate changes (means and extremes) that urban city planners or managers might identify as being of greatest concern (see Table 2). The project team then identified the city services that would be exposed to each climate change effect, as well as the city components (sectors/planning processes) that might be sensitive to those exposures. The combination of these two factors provided us with the areas that need exposure and sensitivity indicators.

For example, Section 2.2.1 focuses on drought as the climate stressor. The project team used peer-reviewed scientific literature on drought resilience (as an example of resilience to a particular effect of climate change) to identify qualitative indicators (i.e., questions for city managers to collect information based on their experience) that help assess a city's capability to reduce exposure and sensitivity to drought, respond to the risks drought poses, and promote learning from previous experiences with drought (see Table 3). The combination of scores for the qualitative indicators and the quantitative indicators of exposure and sensitivity provides a measure of a community's overall resilience to climate events such as drought. In addition to the resilience scores for the indicators, indicators are scored for importance, to reflect the degree to which they contribute to resilience, acknowledging that some indicators reflect issues of higher priority or more direct relevance to urban resilience than others.

As discussed previously, the conceptual framework (see Figure 1) includes the three elements of urban vulnerability (exposure, sensitivity, and response capacity) across any given sector, as well as the process of initiating responsive action, learning, and building risk reduction capacity. This cycle is supported or impacted by the presence of bridges to action, barriers to learning, and barriers to responding.

EPA established a multisector Technical Steering Committee (TSC) to support the development and implementation of an urban climate resilience tool, using the conceptual framework as a foundation (see Appendix A for a list of TSC members). TSC members were selected from local, state, and federal government agencies; academic institutions; nonprofit research institutions or think tanks; and other venues. These individuals came from disciplines that represented different aspects of planning and management relevant to an urban setting and to the eight municipal management sectors within the tool: water, energy, transportation, people (public health and emergency response), economy, land use/land cover, the natural environment, and telecommunications. Each TSC member was assigned to one or more sector subcommittees based on the relevance of his or her background to those sectors.

The results of the TSC's work to select qualitative and quantitative metrics formed the basis of the urban climate resilience tool. This tool uses those indicators (see Appendices G and H for qualitative and quantitative indicators, respectively), along with threshold values for each quantitative indicator for the eight city sectors mentioned above. To apply the tool, local government officials select indicators relevant to their community, evaluate each indicator's importance for representing resilience, score qualitative indicators, and evaluate data results to score quantitative indicators of their community's resilience to climate change. The process of developing this tool is described in more detail in the sections that follow.

2.1. MULTICRITERIA ASSESSMENT AND MIXED METHODS

The project team, in consultation with the TSC, concluded that the analysis of a combination of quantitative indicator data and more subjective scores for qualitative indicators was the assessment approach most likely to be valuable to cities. Such an approach makes use of detailed data sets when they are available, but recognizes that important elements of a city's resilience would be neglected if more subjective information provided by city planners or managers were excluded. Using similar methods and scales from these tool components was critical to developing a set of unified, comparable outputs for analysis. To evaluate how best to integrate these different types of data into a meaningful interpretation of resilience at the city scale, the project team conducted a literature review on methodologies that combine quantitative and qualitative information, with a focus on two areas: mixed methods and multicriteria approaches.

Mixed-methods research is positioned between the quantitative research and qualitative research paradigms, as it synthesizes viewpoints and methods from both. The advantages of mixed-methods research are that it can: provide stronger evidence for a conclusion through the convergence of qualitative and quantitative findings; lead to formulating and answering a broader range of research questions than a single method can; balance the strengths and weaknesses of differing methods; and increase the generalizability of a study's results (Johnson and Onwuegbuzie, 2004).

Multicriteria analysis or multicriteria assessment (MCA) is a set of decision support methods that seeks to select one or a few preferred alternatives based on multiple criteria or objectives (UNFCCC, 2005). MCA studies involving participant engagement (as our study does) solicit input on preferences that is often converted to quantitative data on ordinal scales. In some studies, the information gathered through interviews is exclusively qualitative (De Marchi et al., 2000; Mendoza and Martins, 2006; Scolobig et al., 2008). In other studies, information from participants is complemented by more definitively quantitative data (Scolobig et al., 2008) or surveys (De Marchi et al., 2000; Scolobig et al., 2008). Because these types of MCA studies combine quantitative and qualitative approaches, they are a subset of mixed-methods research that facilitates stakeholders' or decision makers' selection of alternatives or criteria.

The project team developed the tool for the urban resilience case studies based on the approaches taken by Hajkowicz (2008) and the Global Environment Facility (GEF; 2010) (see Appendices D and E for Washington, DC and Worcester, MA case study results, and see Appendix F for comparison of these results). Hajkowicz (2008) used a multicriteria analysis method that included a priority matrix in which study participants ranked the issues presented according to the issue's importance to the participant. GEF (2010) is a more general mixed-methods approach in which each indicator in the assessment was assigned a set of choices that provided a quantitative rating (0 to 3) for that indicator. These studies offered the most practicable approaches for working with indicators of resilience (when hard data were available), while addressing additional relevant issues via expert input from multiple individuals and providing ways to handle both types of information similarly. The tool was designed for a single respondent per sector, preferably the manager for that sector within city government. Consensus among stakeholders was not a design requirement. This approach reduces time and costs, and it targets the tool at those with the greatest power to implement and absorb tool findings.

The project team and sector subcommittees selected the quantitative and qualitative indicators for the tool based on expert knowledge and the literature on climate change and urban resilience. For each qualitative indicator (question), the project team developed four scores (answers) ranging from least resilient to most resilient (see example in Table 4). The project team identified and gathered data for the quantitative indicators (see example in Table 4). Indicators that are related are grouped together with a single indicator from that group designated as a Primary Indicator and the remaining designated as Secondary Indicators. Groupings were developed to assist cities in ensuring that they provide as comprehensive information as possible: When data for primary indicators are not available, one or more secondary indicators from the same group can be considered a reasonable replacement for the missing information; when data for secondary indicators are not available, a primary indicator will certainly suffice. Complete sets of the qualitative and quantitative indicators for the tool are presented by sector in Appendices G and H.

For both the qualitative and quantitative indicators, the project team asked participants to assign an importance weight of 1 through 4. A weight of 1 indicates low importance, and a weight of 4 indicates high importance. For the qualitative indicators, the project team developed four possible ratings, with each indicator corresponding to a resilience score of 1 through 4 (again with 1 representing low resilience and 4 representing high resilience). To score the quantitative indicators, the project team applied four quantitative ranges to the data associated with each indicator (see Section 2.3 for additional information). These ranges correspond with a resilience score of 1 through 4. Participants then selected the scores for the qualitative indicators and reviewed the quantitative indicator ranges to determine the resilience scores. Resilience scores for indicators, sectors, or the city as a whole are best used for comparison over time within the same city. Qualitative and quantitative indicators with high importance weights and high resilience scores demonstrate where cities are most resilient overall. Qualitative and quantitative indicators with high importance weights and low resilience scores demonstrate where cities are least resilient. Areas of city performance with these combinations of rankings are the most critical areas of focus for cities to address as soon as possible. Quadrant plots are used to emphasize results that have this importance/resilience combination. (See Figure 2 in Chapter 3, Section 3.1 for sample quadrant plot and Figures 5 and 6 in Appendix F for quadrant plots populated with data.)

2.2. QUALITATIVE AND QUANTITATIVE INDICATOR DEVELOPMENT

The project team met with each of the sector subcommittees twice to develop the sector-specific qualitative and quantitative indicators for the tool mentioned in the section above. The qualitative indicators provided a way to obtain information for which no quantitative indicator data were available.

2.2.1. Qualitative Indicator Development

The TSC developed a four-step process to establish qualitative indicators (i.e., questions) best suited to determine climate resilience. The final qualitative indicators address all relevant climate stressors and attempt to assess resilience as comprehensively as possible across all sectors (see Appendix B for the full list of qualitative indicators). To demonstrate the process developed and used by the TSC, the sections below lay out the process using drought as an example stressor and water as an example sector. In practice, the TSC repeated the process for all relevant climate stressors across all sectors to develop the final list of questions as qualitative indicators.

2.2.1.1. *Step 1: Identify Climatic Changes/Events of Concern.*

Table 2 is an overview of all potential climate changes that the TSC considered for their potential to affect urban areas. These correspond to the climate stressors referred to in Figure 1. Stakeholders would select stressors that are of greatest concern for their urban area. Assessments of resilience frameworks have suggested that it is critical to distinguish between short- and long-term changes (i.e., extreme events vs. prolonged climate change). Furthermore, the most effective methods of improving resilience target long-term change, but also address some immediate concerns (Engle et al., 2014).

Table 2. Potential climate changes of concern for urban areas

	Wind	Temperature	Precipitation	Sea level rise
Gradual change	± Mean maximum speed ± Strong winds	± Average annual ± Seasonal average ± Daily min and max	± Average annual ± Season average ± Event magnitude/ duration ± Time between events	+ Sea level + Coastal high water
Extreme events	Heat wave (magnitude/duration)			
	+ Storm surge and flooding			
	Droughts (intensity/duration)			
	Floods (magnitude/frequency)			
	Hurricanes (intensity/frequency)			

In Steps 2 through 4, the TSC evaluated and selected indicators for each component of the framework to assess resilience. These steps were repeated for each climatic event or change of concern.

2.2.1.2. *Step 2: Discuss Related Climate Stressors.*

For the purposes of drought, the TSC evaluated the following:

- Changes in the timing, form, or amount of precipitation that favor more frequent or prolonged drought events
- Increased temperature (increased evapotranspiration)
- Increased wind (increased evapotranspiration)

2.2.1.3. *Step 3: Discuss Urban Services Potentially Exposed to Drought and Urban Sectors Potentially Responsible for Managing the Sensitivities of These Services.*

Under this step, the TSC identified (a) urban services potentially exposed to drought that have the potential to affect urban resilience and (b) the urban sectors responsible for managing potential sensitivities of services to drought. This step corresponds to the "exposure" and "sensitivity" elements in Figure 1 that help determine urban vulnerability. Example urban services potentially exposed to drought include the following:

- Water quality
- Groundwater supply
- Surface water supply
- Aquatic habitats, plants, and animals
- Terrestrial habitats, plants, and animals
- Recreational opportunities
- Look and feel of the landscape
- Energy supplied by hydropower, thermoelectric, or nuclear sources

2.2.1.4. *Step 4: Evaluate the Ability to Reduce Exposure/Sensitivity, Enhance Response Capacity, and Learn.*

The final step of this exercise is similar to Step 3. The TSC discussed the urban services exposed to drought (corresponding to the "response" section of Figure 1) and developed a series of questions to help determine a city's ability to (a) reduce exposure or sensitivity, (b) increase response capacity, and (c) learn from past and future experiences with drought. Risk reduction capacity encompasses (a), (b), and (c). In this project, these concepts also compose the role of governance in urban climate resilience.

Sample questions relevant to the water sector are shown for illustrative purposes in Table 3. The questions are based on what Baker et al. (2009) define as the characteristics of an urban area that determine resilience to drought: (a) current condition of the hydrologic environment (both aquifers and water infrastructure), (b) the match between the scale of water governance and the physical (hydrologic) scale in time and space, and (c) the government's capacity to adapt to hydrologic change (administrative and financial capacity to respond). Questions were selected to measure the capacity to reduce risk, recover from drought, and learn to improve future resilience. While the questions in Table 3 are relevant only to the water sector, questions were developed as qualitative indicators for each sector to evaluate how that sector responded to drought. The final qualitative indicators address all areas of concern related to climate and resilience across all sectors. A similar approach was taken for all of the climate changes of concern and exposed services using available literature and input from the TSC Table 3. Water sector questions related to drought sensitivity, response, and learning.

Table 3. Water sector questions related to drought sensitivity, response, and learning.

	Exposure/sensitivity	Increase response capacity	Learning related to drought
Water quality	• Are there water bodies at risk from water pollution during drought?	• Are there mechanisms in place to reduce pollution to at-risk streams during drought? • Are there means of enhancing recovery of water quality following drought, and are those methods ready to implement?	• Is there monitoring to assess the effectiveness of pollution reduction and recovery strategies and means to incorporate that information into management planning? • Are there any barriers to responding to or learning from past or future drought events?
Groundwater supply	• Is the condition of aquifers and water infrastructure adequate to address long-term drought? • Is the condition of aquifers and water infrastructure adequate to address changes in long-term drought risk (duration, frequency, severity)?	• Are there options available to improve the condition of aquifers and water infrastructure? • Do you have local control of your water source(s) or are they managed by an outside entity (private company, another state, etc.)? • Is resource control centralized or distributed? • Is there a joint institutional mechanism through which water can be managed with partners? • Does the joint institutional partnership provide for flexibility to adjust management in the face of extreme events? • Do water allocation laws (e.g., prior appropriations doctrine) limit control of water management? • Is water infrastructure and supply monitored with respect to demand and distribution? • Does the government allow for civic engagement in resource management decision making? • Do mechanisms exist to generate funding for actions that improve resource management?	• Is there a mechanism in place to learn from failures to execute drought response plans for water supplies? • Do management entities regularly evaluate management plans? • Have there ever been adjustments made to management practices in response to evaluations of past drought responses? • Does the evaluation include the assessment of potential future climate change stressors? • Does the capacity exist to access and assess monitoring data?

	Exposure/sensitivity	Increase response capacity	Learning related to drought
Surface water supply		• Do you have local control of your water source(s) or are they managed by an outside entity (private company, another state, etc.)? • Is resource control centralized or distributed? • Is there a joint institutional mechanism through which water can be managed with partners? • Does the joint institutional partnership provide for flexibility to adjust management in the face of extreme events? • Do water allocation laws (e.g., prior appropriations doctrine) limit control of water management? • Is water infrastructure and supply monitored with respect to demand and distribution? • Does government allow for civic engagement in resource management decision making? • Do mechanisms exist to generate funding for actions that improve resource management?	• Do management entities regularly evaluate management plans? • Have there ever been adjustments made to management practices in response to evaluations of past drought responses? • Does the evaluation include the assessment of potential future climate change stressors? • Does the capacity exist to access and assess monitoring data?
Recreation	• Do local water management plans include provisions for local parks and open space? • Will drought have long-term impacts on local parks and open space?	• Is open space used as an adaptation option for protecting water resources during drought?	

2.2.2. Quantitative Indicator Selection

To organize and obtain detailed data sets relevant to urban resilience, the project team created a database of more than 1,400 indicators or metrics derived from the literature on climate change and urban resilience. From this list, specific indicators were selected during meetings with the subcommittees (see example indicator provided in Table 4 and Appendix C for the full list of quantitative indicators).

Table 4. Example qualitative and quantitative indicator from urban resilience tool

a. Example qualitative indicator

Sector	ID#	Question	Score = 4 (highest resilience)	Score = 3	Score = 2	Score = 1 (lowest resilience)
Economy	1	Is the economy of the urban area largely independent, or is it largely dependent on economic activity in other urban areas?	Largely independent	Somewhat independent	Somewhat dependent	Largely dependent

b. Example quantitative indicator

Sector	ID#	Indicator	Definition	Value
Economy	1437	Percentage of city area in 500-year floodplain	This indicator reflects the percentage of the metropolitan area that lies within the 500-year floodplain.	11.0%

For each of the quantitative indicators, threshold values were established defining the upper and lower boundaries of the four resilience categories. Initial thresholds were established through a review of published academic literature, panel data, case studies, and other reports. The thresholds were later calibrated through discussions with expert stakeholders in each case study city (Washington, DC and Worcester, MA). The initial thresholds were designed to represent resilience levels across all U.S. cities. The literature review and threshold development for each indicator followed a stepwise approach. An initial effort was made to identify published analyses for U.S. cities describing categories of resilience with quantitative thresholds for the indicator. If no analyses were available, an attempt was made to identify theoretical resilience thresholds (presumably applicable to any site) based on modeling efforts.

Where such studies were not available, panel data for U.S. cities were examined to establish a range of values for the indicator across the sampled cities, and published literature (academic literature, news articles, etc.) was consulted to determine the indicator's resilience levels for those cities. This step involved triangulating multiple qualitative assessments, which included interpreting discursive regimes, to establish levels of resilience (Olsen, 2004). If data for the indicator were not available for a sample of U.S. cities, case studies from one or several cities were analyzed to determine the level of resilience for those cases, and efforts were made to determine how representative the case was of all U.S. cities, in terms of resilience (Walker, 2006). Finally, if data or case studies were not available for cities, efforts were made to identify state-level data or case studies from which resilience categories were established using the same qualitative triangulation approach, and considering how resilience for the indicator might differ between the state and city level.

2.3. EXAMPLES OF THRESHOLDS FROM PEER-REVIEWED LITERATURE

Two examples of thresholds found in the literature are indicator #460 (Macroinvertebrate Index of Biotic Condition) and indicator #1440 (Palmer Drought Severity Index). Thresholds for indicator #460 are adapted from Weigel et al. (2002). The original five thresholds and those adapted to reflect a resilience score of 1 to 4 are listed in Table 5.

Table 5. Macroinvertebrate Index of Biotic Condition thresholds

Weigel et al. (2002) thresholds	Adapted thresholds	Resilience score
75 to 80 = Very good biotic condition	Greater than 75 = very good biotic condition	Resilience score = 4
60 to 70 = Good biotic condition	56 to 75 = Good biotic condition	Resilience score = 3
50 to 55 = Fair biotic condition	46 to 55 = Fair biotic condition	Resilience score = 2
25 to 45 = Poor biotic condition 0 to 20 = Very poor biotic condition	0 to 45 = Poor or very poor biotic condition	Resilience score = 1

Indicator #1440 (Palmer Drought Severity Index) also uses thresholds adapted from a literature source (Alley, 1984). The original 11 thresholds and those adapted to reflect a resilience score of 1 to 4 are listed in Table 6.

Table 6. Palmer Drought Severity Index (PDSI) thresholds

Alley, 1984 thresholds	Adapted thresholds	Resilience score
≥ 4.00 = Extremely wet	Greater than or equal to −1.99 = Mild drought or no drought	4
3.00 to 3.99 = Very wet		
2.00 to 2.99 = Moderately wet		
1.00 to 1.99 = Slightly wet		
0.50 to 0.99 = Incipient wet spell		
−0.49 to 0.49 = Near normal		
−0.99 to −0.50 = Incipient drought		
−1.99 to −1.00 = Mild drought		
−2.99 to −2.00 = Moderate drought	−2.99 to −2.00 = Moderate drought	3
−3.99 to −3.00 = Severe drought	−3.99 to −3.00 = Severe drought	2
≤ −4.00 = Extreme drought	Less than or equal to −4.00 = Extreme drought	1

2.4. EXAMPLES OF THRESHOLDS FROM GOVERNMENT ORGANIZATIONS

Thresholds for indicator #284 (Physical Habitat Index [PHI]) are drawn from a set of resource briefs prepared by the U.S. National Park Service detailing research on the physical habitat conditions of streams in the National Capital Region Network (Northrup, 2013). The original four thresholds for PHI are listed in Table 7. No changes were needed for the thresholds to correspond with resilience scores of 1 to 4.

Table 7. Physical Habitat Index thresholds

Northrup (2013) thresholds	Thresholds for tool	Resilience score
81 to 100 = Minimally degraded	81 to 100 = Minimally degraded	4
66 to 80 = Partially degraded	66 to 80 = Partially degraded	3
51 to 65 = Degraded	51 to 65 = Degraded	2
0 to 50 = Severely degraded	0 to 50 = Severely degraded	1

2.5. EXAMPLES OF USING QUARTILES TO ASSIGN THRESHOLDS

Thresholds could not be found in the literature for several indicators, so quartiles in the data sets were used as the thresholds for these. Two examples are indicator #1003 (mobility management) and indicator #1396 (percent access to transportation stops). For indicator #1003, the data set was from the Urban Mobility Report produced by the Texas A&M Transportation Institute (Schrank et al., 2012). This report contains operational cost savings for traffic congestion for 101 urban areas in the United States. Each urban area has a per capita operational cost savings value. Thresholds for this indicator were defined as the quartiles of this per capita yearly congestion cost savings data set. These thresholds are listed in Table 8.

Table 8. Mobility management (yearly congestion costs saved by operational treatments per capita) thresholds and scores

Thresholds	Resilience score
Greater than or equal to $32 per person	4
$18 to less than $32 per person	3
$10 to less than $18 per person	2
$2 to less than $10 per person	1

Another example of using quartiles to define thresholds among resilience categories is indicator #1396 (percentage access to transportation stops). Tomer et al. (2011) of the Brookings Institution Metropolitan Policy Program detailed transit accessibility for 100 U.S. cities. Each of the cities in this report contained a value for "share (percentage) of working-age residents near a transit stop." Thresholds were defined as the quartiles of values for the 100 cities in the report. These thresholds are listed in Table 9.

Table 9. Percent access to transportation stops thresholds and scores

Thresholds	Resilience score
76 to 100% of population near a transit stop	4
64 to 75% of population near a transit stop	3
48 to 63% of population near a transit stop	2
23 to 47% of population near a transit stop	1

2.6. DATA COLLECTION APPROACH

The project team designed the data collection approach for the two case studies based on resources and data availability in Worcester, MA and Washington, DC. For Worcester, the tool was used as designed; data were collected (via qualitative and quantitative indicators) for each sector through a series of discussions with the key city personnel responsible for the sector. For the District, the project team convened two workshops to provide input on the tool (including input on individual qualitative and quantitative indicators) and to provide data for the qualitative and quantitative indicators. This process was modified slightly to better reflect a workshop approach, although ultimately one key District representative for each sector scored each qualitative indicator. Additional details on the data collection approaches for Washington, DC and Worcester are included in Appendices D and E, respectively.

3. DISCUSSION AND CONCLUSIONS

This project resulted in a comprehensive tool for identifying the greatest risks, successes, and priorities related to urban vulnerability and resilience to climate change. This effort used the conceptual framework presented in Figure 1 (see Chapter 1) as a foundation. The results can easily be analyzed with respect to exposure/sensitivity, response capacity, or learning, as the qualitative and quantitative indicators are characterized accordingly. In addition, the data collected may be analyzed in the context of the framework, for the purposes of identifying and prioritizing adaptation activities. This prioritization process may involve categorizing critical vulnerabilities (i.e., sectors and issues within sectors for which resilience is low but importance is high) into issues that can be addressed in a straightforward manner with adaptation planning and implementation, versus those over which there is limited or no control.

3.1. VISUALIZING RESILIENCE

Quadrant plots (see Figure 2) were used to visualize the data collected for each case study (see Appendices D and E), and for comparisons across the two case studies (see Appendix F). The quadrants are defined by the combination of resilience and importance scores (1 through 4), and categorized based on priority into the following groups:

- Low priority = high resilience (3 or 4) and low importance (1 or 2)
- Small problems that can add up = resilience and importance both low (1 or 2)
- Monitor for changes = resilience and importance both high (3 or 4)
- Vulnerabilities to address = low resilience (1 or 2) and high importance (3 or 4)

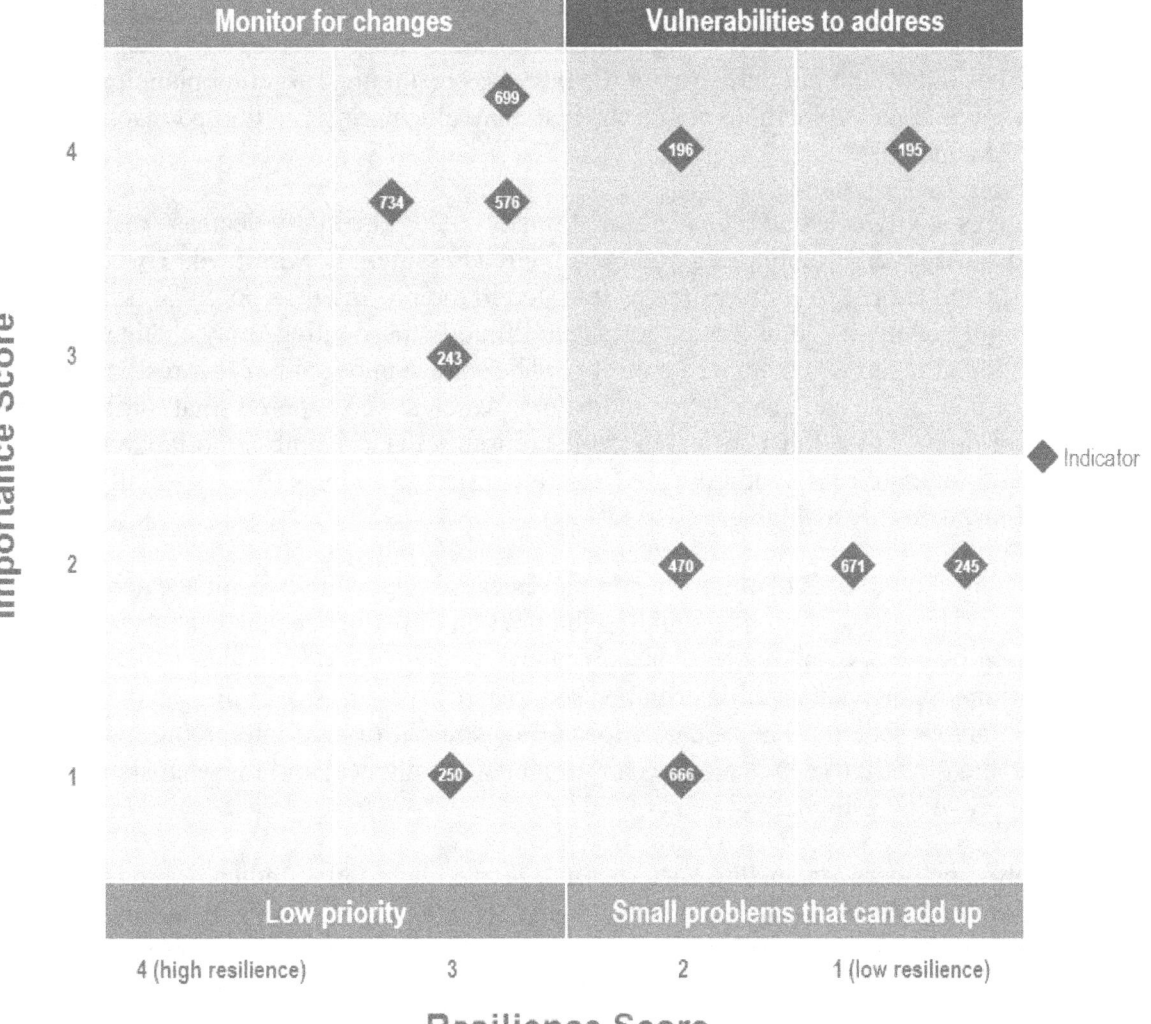

Figure 2. Sample quadrant plot.

These graphics facilitate the interpretation of case study results and are intended to assist city managers in moving on to implement climate change adaptation activities. For example, if a qualitative or quantitative indicator ranked as highly important is also identified as demonstrating high resilience, the city may be considered resilient with respect to that data point or topic ("monitor for changes"), meaning the city is either inherently resilient or has already taken steps to increase resilience. For example, Washington, DC received high resilience and importance ratings for:

- Thermal stress (quantitative indicator)
- Extent to which green infrastructure was selected to provide the maximum ecological benefits (qualitative indicator)

- Extent of telecommunications redundancy and availability of multiple communication options, served by different infrastructure, for first responders and the public (qualitative indicator)

By identifying areas where resilience is high, cities may apply lessons learned to other areas that are also ranked as important, but perhaps significantly less resilient. Targeting planning efforts at a sector's important and vulnerable points can also help cities prioritize limited resources for areas of greatest concern.

However, changes in city characteristics or climate risks could potentially decrease resilience in the future, and some level of monitoring and eventual reassessment is warranted. By contrast, a city can identify and choose to limit resources invested in monitoring data points or issues identified as highly resilient and of low importance (although these ratings may change over time, and the indicator cannot be ignored entirely). The same can be said of low resilience and low importance qualitative and quantitative indicators, which are considered small problems that can add up over time. At the most critical extreme are issues that are both important and have low resilience ("vulnerabilities to address"). These are issues to address first, especially in cases where limited resources are available.

As noted previously, the tool is distinguished in part because it considers resilience across multiple sectors. This allows for understanding the breadth of resilience across a city, relative resilience among its sectors, and the resilience of interdependent sectors, such as water and energy. Additionally, the visualizations can be used to assess progress over time as the tool is used iteratively across sectors. The visualizations allow straightforward interpretations of what the qualitative and quantitative indicators mean for a city's resilience, and for what steps a city may take to improve its resilience.

The identification numbers assigned to each qualitative and quantitative indicator are included in the visualizations to allow the reader to determine exactly what aspects of resilience are being addressed within each quadrant. This is particularly easy in unique situations, for example when a quadrant has few qualitative indicators populating it or few qualitative indicators from a specific sector (even if many from other sectors). The ability to drill down into the data may also be useful for testing hypotheses such as the interrelatedness of certain sectors and their aspects (e.g., do qualitative indicators for sectors that are presumed to be interrelated often fall into the same quadrants, at least for aspects that are presumably interrelated?).

More work is necessary to capture interdependencies among sectors. Future advancements in our understanding of these interdependencies can be made by examining linkages more closely, such as those between the water and the energy sectors. However, interdependencies have been addressed to some extent, in that some qualitative and quantitative indicators have been assigned to more than one sector (e.g., indicator #680: ecological connectivity is used in both the land use/land cover and the natural environment sectors).

3.2. THE UTILITY OF QUALITATIVE ANALYSIS

One of the most significant results of the case studies, and one that was anticipated by the TSC when developing the qualitative indicators and selecting the quantitative indicators, is that the qualitative indicators were found to be essential to the analysis, addressing data quality and availability limitations at both the city and sector level in the two case study applications. While Washington, DC and Worcester, MA contrast each other in data richness and institutional support for climate change adaptation, the project team encountered data sufficiency and availability challenges with both. For example, data were available for only two of the quantitative indicators relevant to the telecommunications sector in Washington, DC, and no data were available for this sector in Worcester. Thus, little to nothing could be understood about the resilience of this sector were it not for the qualitative indicators. The resilience of this sector, however, is critical for responding to extreme climate events. More broadly, the insufficiency of available indicator data was a challenge across almost all sectors in Worcester. Very little could be known about Worcester's resilience from the quantitative indicators alone. The qualitative indicators, however, provided a fairly robust assessment of where Worcester's strengths and weaknesses lay and what issues city managers considered important or more ancillary.

As noted previously, the tool and its conceptual framework reflect the inherently evolving nature of resilience, and the tool's accessibility to the many small- and medium-sized cities across the United States (and internationally) is enhanced by the fact that using tool does not necessarily require data (i.e., if the user applies only the qualitative indicators). The qualitative indicators can be mapped to specific events or types of events, allowing city managers and planners to identify feedbacks and learn over time. As the framework is applied iteratively, qualitative indicators can be reframed to identify specific factors that increase or decrease resilience relative to the previous application of the framework. Additionally, applying the qualitative indicators necessitates interaction with and between sector stakeholders. These interactions provide additional learning and coordination opportunities that would not have been possible using quantitative indicators alone, and these interactions can be used to further refine the resilience assessments and prioritization of activities in response to the assessments' findings. Furthermore, reviews of existing resilience frameworks (e.g., Schipper and Langston, 2014; Engle et al., 2014) have suggested that quantitative indicators should not be used without context, as the value of a single indicator can vary significantly in time and space. Individual quantitative indicators may not appear relevant or may be misleading, unless supplemented by qualitative, contextual information, particularly for local or regional assessments.

Moving forward, it would be wise to evaluate a city's data availability before beginning the more detailed assessment. If data availability is minimal, moving forward with only qualitative indicators may be the most useful approach for evaluating the city's resilience.

3.3. INDICATORS OF RESILIENCE AND THEIR THRESHOLDS

Quantitative indicators were incorporated into the tool to make the best possible use of data on resilience when it was available. Thresholds for the quantitative indicators were based on the literature when possible, accounting for the full range of values the indicator takes on in cities across the United States. Thresholds may need to be reevaluated if applying the tool internationally.

Ideally, thresholds make the resilience assessment more informative because they bring a degree of objectivity to the assessment that does not depend on comparisons with other cities (i.e., this is not a relative resilience assessment). Yet thresholds make use of other cities' experiences in what values the indicators may have had when passing from resilient to less resilient, or vice versa. Indicator thresholds can guide city decisions regarding adaptation (e.g., what to do, how much to do), as the city attempts to avoid exceeding threshold levels that would indicate moving into a state of lowered resilience. Understanding thresholds can increase management efficiency, can help cities prioritize management goals more accurately, and ultimately, may increase the likelihood of a city achieving management targets in key areas.

Although the qualitative indicators lack some of the objectivity provided by the indicator data, they fill in gaps on issues that the quantitative indicators cannot address, often due to data availability limitations, and sometimes because it is impossible to develop an indicator that provides more objective information than the city managers' responses regarding a specific question or issue. However, as the technology enables faster, more efficient, and less expensive data gathering, and citizen-science efforts advance, issues in the tool that are currently addressed only by qualitative indicators might eventually be addressed by quantitative indicators. As it stands now, although, the disparities in available data from the literature and through data collection between the two cities selected for case studies may not only indicate that analysis efforts in cities facing similar issues with data may be difficult, but also the absence of such data in itself indicates potential vulnerabilities to climate change.

The initial application of the tool in a given city, as detailed in the case studies in the appendices, is merely a snapshot of a city's resilience at a given time. The more important evaluation occurs over time, when the tool is applied iteratively to a given city, and can thus better measure learning, increases or decreases in overall resilience, increases or decreases in specific aspects of resilience, and other changes in the urban system. Application of the tool over time can facilitate a city's learning and therefore increase a city's resilience. Resilience is dynamic and evolutionary, and the framework used to develop this tool ensures that the dynamic nature would be built in, allowing the tool to evolve through iterative application.

3.4. SPECIFIC CHALLENGES IDENTIFIED THROUGH TOOL APPLICATION

Beyond the numeric values of resilience and importance collected across the sectors during the case studies (and the supporting data or responses that contributed to those scores), this effort collected important information regarding the challenges that emerged in identifying and confirming appropriate and relevant data sources to effectively assess the proposed indicators. The following discussions identify and expand on previously mentioned challenges encountered in developing and applying the tool.

3.4.1. Discussions with Experts and Gathering City-Specific Knowledge

A major challenge encountered in applying the tool was gathering city-specific knowledge. Different methods were attempted in the two case studies in this report: a workshop approach in Washington, DC and one-on-one discussions in Worcester, MA. The results presented in Appendix D (the Washington, DC case study) reflect the workshop approach, while the results presented in Appendix E (the Worcester, MA case study) represent the responses of the local government official selected as the most knowledgeable in the area addressed by each indicator.

The results are meaningful and can provide city planners and managers with a starting point for prioritizing climate change adaptation activities.

The opinion of one selected expert may not reflect the reality of the situation. There is also an obvious tradeoff: more discussions equate to more time and expense. However, the workshop approach may not necessarily be an improvement because more participants may not mean more viewpoints; groupthink may be an issue, especially if many different representatives of the same agency give their opinion. It is possible that in a workshop setting, lower ranking members are afraid to contradict their superiors. To complicate the issue further, a workshop approach is not possible for many small cities because there may be no more than one or two people with sufficient local expertise to respond with meaningful data.

Ultimately, the value of the information depends on the context in which it is used. For example, findings based on a limited set of consultations may not be immediately actionable, but they may be valuable in raising awareness of issues where more study is needed and the expense of broadening data collection may be justified. Limited data collection may also be more useful in an aggregate sense; while there may be little to learn about a specific item, there may be greater confidence about the average state of the sector. Broadly, we may learn about the average of the sectors overall; even if a sector is only represented by a single opinion, there can be overall confidence that a city needs to approach its climate resilience planning more seriously in that area. In addition, supplemental indicator data help balance out any subjectivity that could influence responses or importance rankings for both qualitative and quantitative indicators.

3.4.2. Lack of Data and Spatial/Temporal Data Variability

Lack of quantitative data was a significant limitation, particularly in the Worcester, MA case study. Table 10 below highlights the lack of data for many sectors; no data were available for two sectors, and all desired data were available for only one sector. Based on the results of the case studies, it is likely that this problem is systemic to many small- and medium-sized cities.

Table 10. Worcester, MA data availability

Sector	Percentage of total indicators with quantitative data
Economy	100%
Land use/land change	58%
People	50%
Water	50%
Transportation	42%
Natural environment	10%
Energy	0%
Telecommunications	0%

Data availability problems were not confined to Worcester; data availability in the District was also limited in some cases. One advantage unique to the District was that data for several indicators were available from sources maintained by District government entities and national databases on U.S. cities or states, where the District had the advantage of being treated as both a state and a city.

However, data availability was still the most significant limitation in applying the urban resilience tool to evaluate the District's resilience. Although some data were typically available, many issues with data quality were observed. Factors characterizing data limitations are listed in Table 11. Regardless of the city size, data limitations will continue to be an issue when using this tool. However, this problem is not unique; nearly all real-world policy tools face data gaps. Instead, it is important to remember that any analysis and findings must consider the limitations specific to the city and sector and properly frame any conclusions or recommendations in light of these limitations.

Table 11. Quantitative data limitations

Data limitation	Description
Data not available	No data were available for the indicator.
Significant postprocessing	Data were available but required significant processing to obtain the value of the indicator for the city.
Multiple data sets	Calculating the indicator value required more than one data set; in some cases, combining data sets was challenging due to different spatial and temporal resolutions.
Modeled data	Extensive modeling efforts were required to calculate the indicator value.
Ongoing data collection	Data collection efforts were proposed or ongoing and therefore incomplete.
Outdated data	Available data were out of date and inappropriate for measuring the current resilience of the city.
Regional-scale data	Data were available only at a regional scale (e.g., county), not at the municipal scale.

The spatial variability of data can also be an issue. Within a city and within a sector, service quality and vulnerability may vary, even from block to block, depending on a host of factors (e.g., elevation, maintenance schedule, districting). Aggregating data at a city level may hide problems that are only severe in specific instances, or in the opposite case, make problems that are limited to small areas appear much worse than they are. For example, localized flash flooding may be more important than extreme large-scale events, especially in Worcester, given its hilly topography. Given that the impacts of climate change stress are highly variable in space, a geographic information system (GIS) approach (or other approaches in conjunction with GIS) to mapping potential climate-related impacts across the urban area and its surrounding region could inform resilience assessment and planning efforts.

Lastly, temporal variability may pose additional challenges. Many data are historical, yet climate vulnerabilities are often the result of deviations from the historical pattern. For example, under climate change scenarios, the new 500-year flood area may be considerably larger than the existing/historical 500-year flood area. Another way of looking at this is that the historical 500-year flood is likely to happen more frequently. In the District, rapid gentrification may have rendered historical data sets less informative, even those collected as little as a decade ago. Data sets may be of limited use because they do not reflect the future conditions that would inform planning, or they must be modified (consuming time and expertise) to be useful.

3.4.3. Sector Interconnectivity

This exercise also underscored the interconnectivity of sectors. For example, energy providers are highly dependent on the transportation sector and emergency response during extreme weather events. Likewise, the continued provision of safe drinking water relies heavily on a resilient energy sector, and disruptions to water service could have significant impacts on public health and the economy, among other sectors. Cities may need to dig deeper into a specific sector to understand the true nature of the vulnerability. The interconnected complexity associated with measuring climate-related resilience requires considering each sector as a dynamic system comprised of subsystems. For example, the water/wastewater sector could be divided into eight subsystems: (1) water sources, (2) extraction of source water, (3) water treatment, (4) water distribution to users, (5) wastewater discharge by water users, (6) wastewater (and stormwater) collection, (7) wastewater treatment, and (8) effluent returns to receiving waters and/or water recycling.

This tool also homogenizes some vulnerabilities that occur because of interconnectivity across sectors. Low scores in one sector may be largely due to another connected sector's poor performance. For example, city water experts may rank the resilience of the water sector low because they are aware that power for and transportation to water infrastructure may be difficult or nonexistent in emergencies. In effect, the poor relative performance of a single sector that is interconnected to others may lower scores in many sectors, masking the fact that a single sector is responsible for low resilience across sectors.

3.4.4. Revisions to Qualitative and Quantitative Indicators

Several participants challenged the qualitative and quantitative indicators used. Generally, the participants noted three concerns: (1) the proposed qualitative or quantitative indicator is not assessing what really matters, (2) the proposed qualitative or quantitative indicator is poorly defined, or (3) the proposed quantitative indicator would be very difficult to measure and monitor.

Many of the issues that surfaced showed the deep technical and site-specific experience needed to create and apply meaningful qualitative and quantitative indicators. For example, with regard to quantitative indicator #983 (average customer energy outage [hours] in recent major storm), one participant stated that this indicator attempts to address a relevant metric, but it does not speak to the most important variable: the indicator should account for *when* the power is out, which is more important than the length of time the power is out. A power outage in the middle of the night may have limited effects, and even frequent nighttime outages may not indicate low resilience. However, a power outage in the middle of the day or an outage that occurs during a heat wave or extreme event leaves city residents much more vulnerable. However, the issue of when climate impacts occur is relevant to many of the indicators in the tool, and incorporating timing, or similar details, into every indicator may be unnecessarily complicated for this tool.

These concerns show the need for continued refinement of the tool. At the same time, this also highlights the usefulness of the framework used to develop the tool, which is iterative and evolutionary by design. The qualitative and quantitative indicators selected for the tool are, in many cases, akin to canaries in a coal mine. They provide a sense of what might happen to a city facing threats to its resilience due to climate change. The most important data may be that which shows that a particular sector or subsector is close to reaching a resilience threshold, and that the system might fail if further stressed by future climate change. While iterative application of the tool, and application of the tool across additional urban communities will help refine some qualitative and quantitative indicators, we have attempted to select the most relevant indicators in this first version of the tool—those that best reflect important aspects of resilience and show where a city is solidly resilient or at risk from future stressors.

3.4.5. Threshold-Setting

Thresholds are not easy to set for all indicators. Ideally, the thresholds that determine when the values of an indicator pass from a state of nonresilience to a state of resilience would be objectively determined based on a full understanding of the indicator and its components. However, identifying objective thresholds for indicators is difficult, as they (a) are not widely applicable at different scales; (b) can be challenging to identify in social, environmental, or economic settings unless they have been breached and a disturbance has been observed; and (c) might vary spatially and temporally, making it difficult to apply them uniformly across different cities (U.S. EPA, 2011). For most indicators in this study, objective thresholds were not available in the published literature. When thresholds were available, they varied spatially across the United States, and in some cases, were not applicable to Washington, DC nor Worcester, MA. For this tool, the challenge was resolved by asking the city planners or managers to review draft thresholds and change them if needed based on their expert knowledge. Few thresholds were changed during the first two pilot applications of the tool, as threshold determination is intended to take place primarily during the tool development phase. The intent is that eventually, after additional applications of and refinements to the tool, the thresholds would be unlikely to be changed. Iterative application of the tool over time and in multiple locations can help determine the most accurate thresholds to use; the framework used in this study can evolve as city planners or managers assess resilience and implement action plans.

3.4.6. Integrating Qualitative Information

The urban resilience tool aims to provide a quantitative measure of a city's resilience to climate change. However, to reflect the many urban management areas and issues related to a city's resilience, it is necessary to include information when measured data are not available or an issue does not lend itself to immediate quantitative measurement. Therefore, information on specific metrics that are measured by city departments, government agencies at other levels (e.g., state or federal), or other entities should be supplemented by more subjective information based on managerial experience. Finding a methodology that integrates subjective information from experts and metrics that indicate resilience on a single scale is key. In this tool, the project team used a mixed-methods approach (discussed in Section 2.1) to integrate information obtained from city planners or managers with information obtained through data searches to overcome this challenge.

Integrating quantitative data from resilience indicators with qualitative information from responses to questions was essential. Washington, DC is a city that is more likely than usual to have indicator data available because of its unique relationship with the federal government and because in many ways, it functions as its own state. But even there, subjective information from city planners or managers was needed to address all of the desired facets of climate resilience. Only with a mix of quantitative data and qualitative questions was the tool able to fully capture where cities were or were not climate-resilient. In Worcester, which lacked detailed, up-to-date indicator data, the questions asked of city planners and managers were even more important. In all likelihood, Worcester's data availability is more representative of American cities than the District's. Finding meaningful ways to incorporate expert judgment is essential for any tool that wishes to find widespread use in urban areas across the United States.

Information from city planners and managers also supplemented known data shortcomings. Even where indicator data were available, the information may not reflect the full spatial and temporal diversity of the city. For example, the data set may start or stop before capturing weather events such as floods or droughts, which may occur only at long intervals (e.g., every 100 or 200 years). Additionally, poor sampling in existing studies may fail to include all subpopulations. Information provided by city planners and managers based on their experience allows for these data shortcomings to be partially accounted for by the tool.

The results of the case studies in Washington, DC and Worcester, MA suggest that there may be more utility to focusing on qualitative information, especially in cities with limited data and resources. Furthermore, by applying this tool iteratively to the same city over time, city planners and managers can optimize the collection of qualitative information and potentially identify areas where targeted quantitative data collection would be most beneficial. The efficient allocation of limited resources is critical to increasing urban resilience, as cities are unable to address all issues, and must prioritize their time, effort, and investments appropriately.

3.5. FUTURE STEPS

Urban centers in the United States (and globally) have a long way to go in adapting to climate change. City planners and managers in both Washington, DC and Worcester, MA agreed that the urban resilience assessment tool developed by EPA provided valuable insights into the resilience of their respective cities and assessment results can be incorporated into ongoing planning. In many cases the information provided by the tool yielded as many new questions as answers; however, this aspect of the tool can assist city managers and utilities in identifying further issues to pursue to improve their resilience. In this context, the evolutionary nature of the framework used to develop this tool is particularly important. The tool's greatest utility is in applying it repeatedly over time to the same city to better understand current and future resilience and critically evaluate the successes and failures of adaptation initiatives. With new patterns of more extreme weather across the globe, adaptation is essential. A first step for many cities will be to assess their sector-specific and overall resilience to climate change. Additional steps could include:

- **Additional applications of the tool**
  With a sample size of two, it is difficult to draw conclusions about how the tool will perform across a wider swath of American cities. With additional cities applying the tool to perform assessments in different environments, the tool could be tested in new ways and its applicability and value to a broad range of city governments could be refined. For example, it is not known how well the tool would perform for cities in western or

southern parts of the country; these cities have built environments that are more recent than the early 1700–1800s and face different climate change risks. Likewise, it is not known how well the tool would perform for resource-limited cities with similar risk profiles that pool their resources to address climate vulnerability more efficiently, as some cities and counties are starting to do now.

- **Lesson-sharing and best practices**
Many regions have cities that share common histories and common climate risks. Learning from similar cities will be essential for all cities planning to address climate vulnerabilities. How could the tool help develop successful adaptation strategies, and how could those adaptation strategies be shared with cities that share the same vulnerability profile (a similar set of values across indicators)? Smit and Wandel (2006) showed that community-based adaptation opportunities are multidimensional and affected by exposures, sensitivities, adaptive capacities, and other factors. Application of the tool by more cities can help identify commonalities in these factors and opportunities for sharing best practices, policies, and adaptive strategies more quickly and effectively to meet the growing challenge of overcoming climate risks.

APPENDIX A. TECHNICAL STEERING COMMITTEE MEMBERS

Technical Steering Committee Members

Name	Agency	Expertise	Indicator Subcommittee
Baranowski, Curt	U.S. EPA	Water utilities and security	• Energy and water
Cain, Alexis	U.S. EPA Region 5	U.S. EPA regional science representative	• Natural environment
Carmin, JoAnn	Massachusetts Institute of Technology	Sociology and climate adaptation	• People
Chan, Steve	Harvard University	Information technology	• Telecommunications
Cutter, Susan	University of South Carolina	Hazards and disasters	• Energy and water • Natural environment
Farris, Laura	U.S. EPA Region 8	Engineering and climate change	• Energy and water • Transportation • Natural environment
Fay, Kate	U.S. EPA Region 8	U.S. EPA regional science representative	• Natural environment
Gonzalez, Larry	U.S. EPA Region 7	U.S. EPA regional science representative	• Natural environment
Goold, Megan	U.S. EPA Region 3	U.S. EPA regional science representative	• Natural environment
Greene, Cynthia	U.S. EPA Region 1	U.S. EPA regional science representative	• Natural environment
Gross-Davis, Carol Ann	U.S. EPA Region 3	U.S. EPA regional science representative	• Natural environment • People
Hansen, Verle	U.S. EPA	Land use planning	• Energy and water • Transportation • People
Hodgeson, Kimberley	Cultivating Sustainable Communities	Public health and urban planning	• Energy and water • People
Holway, Jim	Sonoran Institute	Land use and water resources planning and smart growth	• Energy and water • Transportation • Natural environment

Name	Agency	Expertise	Indicator Subcommittee
Hulting, Melissa	U.S. EPA Region 5	U.S. EPA regional science representative	• Natural environment
Jackson, Laura	U.S. EPA	Ecology	• Natural environment
Jencks, Rosey	San Francisco Public Utilities Commission	Stormwater engineering	• Energy and water
Jones, Bill	U.S. EPA Region 3	U.S. EPA regional science representative	• Natural environment
Kafalenos, Robert	United States Department of Transportation (U.S.DOT)	Transportation	• Transportation
Kasperson, Roger	Clark University	Risks and uncertainty	• Energy and water • People
Kreider, Andrew	U.S. EPA Region 3	U.S. EPA regional science representative	• Natural environment
LaGro, James	University of Wisconsin–Madison	Urban planning	• Energy and water • Transportation • Natural environment
Lawson, Linda	U.S.DOT	Transportation	• Transportation
Leichenko, Robin	Rutgers University	Economics and finance	• Economy
Lupes, Rebecca	U.S.DOT	Transportation	• Transportation
Machol, Ben	U.S. EPA Region 9	U.S. EPA regional science representative	• Natural environment
McCullough, Jody	U.S.DOT	Transportation	• Transportation
McGeehin, Michael	Retired	Human health	• People
Mitchell, Ken	U.S. EPA Region 4	U.S. EPA regional science representative	• Natural environment
Narvaez, Madonna	U.S. EPA Region 10	U.S. EPA regional science representative	• Natural environment
Newman, Erin	U.S. EPA Region 5	U.S. EPA regional science representative	• Natural environment
Olson, Kim	U.S. EPA Region 7	U.S. EPA regional science representative	• Natural environment

Name	Agency	Expertise	Indicator Subcommittee
Pincetl, Stephanie	University of California Los Angeles	Urban planning	• Energy and water • Transportation • Natural environment
Pyke, Chris	U.S. Green Building Council	Green building	• Energy and water
Raven, Jeffrey	Architect	Architecture and sustainability	• Energy and water • Transportation • Natural environment • People
Quay, Ray	Decision Center for a Desert City Arizona State University	City planning, water and wastewater	• Land use/land cover • Natural environment
Rimer, Linda	U.S. EPA	Regional and local climate adaptation	• Energy and water • People
Rosenberg, Julie	U.S. EPA	Climate change, mitigation, and cities	• Energy and water • People
Ruth, Matthias	Northeastern University	Governance	• Energy and water • Economy
Rypinski, Art	U.S.DOT	Transportation	• Transportation
Santiago Fink, Helen	U.S. Agency for International Development	Planning and international	• Natural environment • People
Saracino, Ray	U.S. EPA Region 9	U.S. EPA regional science representative	• N/A
Schary, Claire	U.S. EPA Region 10	U.S. EPA regional science representative	• Natural environment
Scheraga, Joel	U.S. EPA	Economics and finance	• Economy
Shephard, Peggy	WE ACT for Environmental Justice	Environmental justice	• Energy and water • Economy • People
Smith, Gavin	University of North Carolina–Chapel Hill	Disasters and hazards	• Energy and water • Natural environment
Solecki, Bill	Hunter College of the City University of New York	Economics and finance	• Economy

Name	Agency	Expertise	Indicator Subcommittee
Spector, Carl	City of Boston	Air quality	• Natural environment
Stults, Missy	National Climate Assessment	Urban sustainability	• Energy and water • Transportation • Natural environment
Susman, Megan	U.S. EPA	Urban planning and smart growth	• Energy and water • Transportation • Natural environment
Wilbanks, Tom	Department of Energy	Energy systems	• Energy and water
Willard, Norman	U.S. EPA Region 1	Climate change and state and regional policy	• Energy and water • People
Wong, Shutsu	U.S. EPA Region 1	U.S. EPA regional science representative	• Natural environment
Yarbrough, James	U.S. EPA Region 6	U.S. EPA regional science representative	• Natural environment
Zinsmeister, Emma	U.S. EPA	Climate change, mitigation, and cities	• Energy and water • People

APPENDIX B. PARTICIPANTS

Worcester, MA Participants

Sector	Participant Name	Participant Title	Organization
Economy	Timothy Murray	President and CEO	Worcester Regional Chamber of Commerce
Energy	John Odell	Worcester Energy Manager	City of Worcester
Land use/land cover	Luba Zhaurova	Acting City Planner	City of Worcester
Natural environment	Rob Antonelli, Jr.	Assistant Commissioner of Parks and Recreation	City of Worcester
People	Derek Brindisi	Director, Worcester Department of Public Health	City of Worcester
	Kerry Clark Seth Peters Colleen Turpin	Worcester Department of Public Health	City of Worcester
Telecommunications	David Clemons	Director of Emergency Communications and Management	City of Worcester
Transportation	Bill Moisuk	Principal Planner (transportation)	Central MA Regional Planning Commission
Water	Konstantin Eliadi	Director, Water and Sewer Operations	City of Worcester
	Phil Guerin	Director of Environmental Systems, Worcester Dept. Public Works	City of Worcester
	Karla Sangey	Director	Upper Blackstone Pollution Abatement District (UBPAD) Treatment Plant, Millbury, MA
	Mark Johnson	Deputy Director	UBPAD Treatment Plant, Millbury, MA

Washington, DC Workshop Participants

Name	Affiliation	Sector	First Workshop	Second Workshop
Brendan Shane	Chief, Office of Policy and Sustainability District Department of the Environment	Cross-cutting*	X	X
Wendy Hado	District Department of the Environment	Cross-cutting*		X
Maribeth DeLorenzo	Sr. Policy Specialist Department of Housing and Community Development	Economy		X
Tanya Stern	Chief of Staff Office of Planning	Economy	X	Sent responses ahead of time
Andrea Limauro	Office of Planning	Economy	X	
Emil King	Policy Analyst District Department of the Environment	Energy	X	X
Jessica Daniels	District Department of the Environment	Energy	X	X
Wesley McNealy	Director, Corporate Environmental Services Pepco Holdings, Inc.	Energy	X	X
Sean Skulley	Sr. Specialist, Sustainability and Business Development Washington Gas	Energy	X	X
Shirley Harmon	Pepco Holdings, Inc.	Energy		X
Susan Nelson	E9-1-1 Coordinator-COOP Coordinator Office of Unified Communications	Telecommunications	X	
Christopher Bennett	IT Program Manager Office of the Chief Technology Officer	Telecommunications	X	X
Laine Cidlowski	Urban Sustainability Planner Office of Planning	Land use/land cover	X	X
Damien Ossi	Wildlife Biologist District Department of the Environment	Natural environment	X	
Rama Tangirala	District Department of the Environment	Natural environment	X	
Dan Guilbeault	Policy Analyst District Department of the Environment	Natural environment		X

Name	Affiliation	Sector	First Workshop	Second Workshop
John Davies-Cole	State Epidemiologist District Department of Health	People	X	X
LaVerne Hawkins Jones	Asthma Control Program Manager Department of Health	People		X
John Thomas	State Forester District Department of Transportation	Transportation		X
Rachel Healy	Sustainability Project Manager Washington Metropolitan Area Transit Authority	Transportation		X
Gregory Vernon	Washington Metropolitan Area Transit Authority	Transportation	X	
Phetmano Phannavong	Environmental Engineer District Department of the Environment	Water	X	
Shabir Choudhary	Section Chief Washington Aqueduct	Water	X	
Maureen Holman	Sustainability Manager DC Water	Water	X	X
Jonathan Reeves	Emergency Response and Planning Coordinator DC Water	Water		X

* Attendees with cross-cutting expertise were asked to select the sector group to which they believed they could contribute the best input.

Workshop Observers*

Name	Affiliation	First Workshop	Second Workshop
Aaron Ray	Associate Georgetown Climate Center	X	
Amy Tarce	Urban planner National Capital Planning Commission	X	X
Ann Kosmal	Convener, GSA Climate Adaptation and Resiliency Team Office of Federal High-Performance Green Buildings General Services Administration	X	X
Emily Seyller	U.S. Global Change Research Program	X	
Gerald (Jerry) Filbin	Office of Policy Coordinator for Climate Change Adaptation Environmental Protection Agency	X	
Jalonne White-Newsome	Federal Policy Analyst West Harlem Environmental Action, Inc. (WE ACT for Environmental Justice)		X
Robin Snyder	General Services Administration		X
Sara Hoverter	Green Committee Georgetown Law	X	
Shana Udvardy	Climate Adaptation Policy Advisor Center for Clean Air Policy	X	

*Observers were asked to select the sector group to which they believed they could contribute the best input.

APPENDIX C. AGENDAS FOR WORKSHOPS IN WASHINGTON, DC

Meeting on
Assessing Urban Resilience in Washington, DC
11:30am–4:00pm
November 18, 2013
District Department of the Environment Offices,
5th Floor, 1200 First Street NE, Washington DC 20002
Organized by
United States Environmental Protection Agency (EPA) Office of Research and Development,
District Department of the Environment (DDOE), and The Cadmus Group, Inc.

11:30am – 12:15pm	Project Background and Introduction
11:30 – 11:45	Welcome; DDOE's climate urban resilience workshops and climate adaptation plan — *Brendan Shane (DDOE)*
11:45 – 12:15	Indicator thresholds and preliminary tool results for DC — *Julie Blue (Cadmus)*
12:15pm – 12:30pm	Break
12:30pm – 2:30pm	Working Lunch and Breakout Sessions: Scoring Questions and Indicators

Water	**Energy**
Led by Laura Dufresne (Cadmus)	*Led by Vanessa Leiby (Cadmus)*
Shabir Choudhary	Jessica Daniels
Jonathan Reeves	Shirley Harmon
Steve Saari	Emil King
[Holly Wootten]	Wesley McNealy
	Sean Skulley
	[Angie Murdukhayeva]

Natural Environment	**Economy**
Led by Nathan Smith (Cadmus)	*Led by Patricia Hertzler (Cadmus)*
Cecily Beall	Maribeth DeLorenzo
Rama Tangirala	Andrea Limauro
[Jenna Tipaldi]	Tanya Stern
	[Tara Fortier]

Transportation	**Land Use/Land Cover**
Led by Damon Fordham (Cadmus)	*Led by Chi Ho Sham (Cadmus)*
Daniel Lee	Laine Cidlowski
[Sarah Yardley]	*[Anna Weber]*
People	**Telecommunications**
Led by Victoria Kiechel (Cadmus)	*Led by Ken Klewicki (Cadmus)*
Victoria Alabi	Christopher Bennett
John Davies-Cole	Donte Lucas
Russell Gardner	*[Ken Klewicki]*
LaVerne Hawkins Jones	
Jamal Jones	
Wes McDermott	
[Kristin Taddei]	

The following attendees may join any sector or move among sectors:

Amanda Campbell, Ann Kosmal, Brendan Shane, Robin Snyder, Amy Tarce, and Jalonne White-Newsome.

2:30pm – 4:00pm	**Debrief and Discussion of Sectors' Contributions to DC's Resilience**
2:30 – 3:30	Debrief on qualitative and quantitative indicators; discussion of sectors' contributions to DC's resilience — *Julie Blue (Cadmus)*
3:30 – 4:00	Closing remarks — *Susan Julius (EPA)*

Meeting on
Assessing Urban Resilience in Washington, DC
9:00am–4:45pm
September 10, 2013
District Department of the Environment Offices,
5th Floor, 1200 First Street NE, Washington DC 20002
Organized by
United States Environmental Protection Agency (EPA) Office of Research and Development, District Department of the Environment (DDOE), and The Cadmus Group, Inc.

9:00am – 10:20am	**Project Background and Introduction to the Scoring Questions Breakout Sessions**
9:00 – 9:40	Welcome and attendee introductions; urban resilience project framework; complementary projects and health work — *Susan Julius (EPA) and John Heermans (DDOE)*
9:40 – 10:00	Background on DC case study — *Nathan Smith (Cadmus)*
10:00 – 10:20	Methodology for urban resilience tool — *Julie Blue (Cadmus)*
10:20am – 10:30am	**Break**
10:30am – 12:45pm	**Breakout Sessions and Lunch: Scoring Questions**

<u>Water</u>	<u>Energy</u>
Led by Tracy Mehan (Cadmus)	*Led by Vanessa Leiby (Cadmus)*
Shabir Choudhary	Wesley McNealy
Maureen Holman	Sean Skulley
[Ken Klewicki]	*[Angie Murdukhayeva]*

<u>Natural Environment</u>	<u>Economy</u>
Led by Nathan Smith (Cadmus)	*Led by Patricia Hertzler (Cadmus)*
Jessica Daniels	Lee Goldstein
Emil King	Tanya Stern
Damien Ossi	*[Tara Fortier]*
Phetmano Phannavong	
Rama Tangirala	
[Jenna Tipaldi]	

Transportation	**Land Use/Land Cover**
Led by Damon Fordham (Cadmus)	*Led by Chi Ho Sham (Cadmus)*
Rachel Healy	Laine Cidlowski
Gregory Vernon	John Thomas
[Sarah Yardley]	[Anna Weber]

Telecommunications	**People**
Led by Holly Wootten (Cadmus)	*Led by Victoria Kiechel (Cadmus)*
Tegene Baharu	Russell Gardner
Chris Bennett	Peggy Keller
Donte Lucas	[Victoria Kiechel]
Susan Nelson	
[Holly Wootten]	

The following attendees may join any sector or move between sectors:

Gerald (Jerry) Filbin, Sara Hoverter, Ann Kosmal, Aaron Ray, Emily Seyller, Brendan Shane, Amy Tarce, and Shana Udvardy.

12:45pm – 1:25 pm	**Adaptation Planning and Introduction to Indicator Breakout Session**
12:45 – 1:05	Adaptation in DC and upcoming adaptation plan — *Clare Stankwitz (Cadmus) and John Heermans (DDOE)*
1:05 – 1:25	Background on indicators and data sources — *Julie Blue (Cadmus)*
1:25pm – 3:15pm	**Breakout Session: Indicators**
	(Same as morning breakout groups)
3:15pm – 4:45pm	**Debrief and Closing**
3:15 – 4:15	Debrief on qualitative and quantitative indicators
4:15 – 4:45	Closing remarks — *Susan Julius (EPA)*

APPENDIX D. WASHINGTON, DC CASE STUDY

This appendix contains the Washington, DC case study. Section D.1 provides background on the known climate vulnerabilities faced by Washington, DC and any existing planning the city has undertaken to address these vulnerabilities. Section D.2 reviews the results for Washington, DC. Results are by sector and accompanied by visual data summaries.

D.1. WASHINGTON, DC Background

Washington, DC (also referred to as DC or the District) is a major and growing East Coast population center that provides an opportunity to test the urban resilience tool in a city with significant planning, financial, and data resources. The District has already begun climate change resilience and adaptation efforts (see Section D.1.1.2), which allows testing the tool in an environment where the results can augment existing or upcoming adaptation planning efforts, including the Climate Change Adaptation Plan under development by the District (DDOT, 2013). The outcome of this effort includes an unprecedented union of expert judgment and quantitative data to assess the District's climate change resilience that is complementary to DC's already extensive ongoing efforts. The combined outcome of these initiatives provide the District with a more nuanced analysis of the areas in which resilience can and should be strengthened. It also supports many of the Sustainable DC[1] initiative's existing economic, environmental, public health, and quality of life goals (Sustainable DC, 2015).

Washington, DC is located on the Atlantic Coastal Plain at the confluence of the Anacostia and Potomac Rivers, which flow into the Chesapeake Bay. At its lowest point along the Potomac River, the District is at sea level. The flat topography of the coastal plain puts the area at high risk from sea level rise and storm surges from hurricanes and other storms.

The District's population (currently over 646,000) grew by an estimated 7.4% between 2010 and 2013 (U.S. Census Bureau, 2013a). The federal government and services provided for it constitute a large portion of the District's economy. Tourism is also a major component of the local economy. These components are reflected in the sectors that employ the greatest numbers of people: professional; scientific and technical services; education; healthcare and social assistance; and public administration (U.S. Census Bureau, 2013b).

Despite a strong, stable economy that has produced a median income approximately 21% above the national median, the District's poverty rates are higher than average (U.S. Census Bureau, 2013a). Therefore, the District's population ranges broadly from those with a great need for more resilience to those who are already highly resilient to impacts. In recent years, the District has rapidly gentrified; home prices in nearly one-fifth of the city's census tracts moved from the bottom half to the top half of overall citywide housing prices over the period 2000–2007. Nationally, this is the fifth highest rate of gentrification (behind Boston, MA; Seattle, WA; New

---

[1] The Sustainable DC planning initiative began in 2011 and is led by the District Department of the Environment and Office of Planning; its goal is to "make DC the most sustainable city in the nation." (Sustainable DC, 2015).

York City, NY; and San Francisco, CA) (Hartley, 2013). As a result, older data sets may not reflect current demographics.

D.1.1. KNOWN VULNERABILITIES

D.1.1.1. *EXTREME WEATHER EVENTS*

The following types of extreme weather events have been identified by public agencies as posing either a "high" or "medium-high" risk to counties in or near the DC metropolitan region. Climate change may exacerbate these events, which include drought, extreme heat, flash/river flooding, thunderstorms, tornadoes, winter weather (ice and snow), and tidal/coastal flooding (MWCOG, 2013c). Hurricanes, thunderstorms, lightning, hail, wind, and tornados are estimated to cost the DC metropolitan region more than $14 million in damages annually (MWCOG, 2013a). Six recent extreme weather events in the District have tested the resilience of the city's institutions and material infrastructure, as shown in Table 12.

D.1.1.2. *TEMPERATURE*

Not only have temperatures in the DC area risen over the past century, the pace of warming has increased (MWCOG, 2008; Kaushal et al., 2010). The District Department of Transportation (DDOT) has identified trees and vegetation as assets that are vulnerable to the effects of rising temperatures (DDOT, 2013). In the coming century, surface air temperatures in the region are projected to rise another 6.5°F (3.6°C; IPCC, 2007b). The District is a documented urban heat island, with its downtown 10–15°F hotter than nearby rural regions on summer afternoons. The number of days "dangerous" to health within city limits has increased from 8–10% of summer days in the 1950s and 1960s to 18% of summer days in the last decade (Kalkstein et al., 2013). Some of these increases could be potentially reversed through adaptation. Modeling suggests minor (10%) increases in reflectivity and vegetative cover would save approximately 20 lives per decade and also reduce the number of heat-related hospital admissions (Kalkstein et al., 2013).

Higher temperatures and the expected changes in rainfall patterns will change the ecological profile (trees and vegetation) of the region. Over time, crop species and forest species currently characteristic of the Mid-Atlantic region (e.g., apples and grapes; maple-beech-birch deciduous forest) might no longer be viable. While overall forest productivity might increase, the increase in temperatures is also likely to result in increased invasive species and reduced biodiversity, as well as more frequent and more severe forest fires (MWCOG, 2008, 2011a, 2013a). The earlier onset of spring resulting from this warming will affect individuals with pollen allergies, as well as the local tourist industry (including the annual Cherry Blossom Festival). The peak bloom date for cherry blossoms could be 5 to 13 days earlier in year 2050 than today (Chung et al., 2011; Abu-Asab et al., 2001).

Threats to DC's infrastructure from higher temperatures include deterioration and buckling of pavement; thermal expansion of joints on bridges; and premature deterioration of buildings, other infrastructure, sealants, and paints. Maintenance requirements for roads, parking lots, and airport runways might be affected (DDOT, 2013; MWCOG, 2011c, 2013a). Buildings and pavement currently cover more than 40%of the District, producing a pronounced urban heat island effect (Chuang and Hoverter, 2012).

Table 12. Major weather events and their impacts in the District of Columbia since 2003

Weather event	Date	Impacts
Hurricane Isabel	September 2003	Approximately 129,000 customers lost power primarily due to fallen trees and strong winds (NOAA, 2008). The Anacostia River surged over the seawall, causing severe damage to 12 National Park Service offices and the U.S. Park Police Anacostia Operations Facility (NCPC, 2008).
Heavy precipitation and flash floods	June 2006	Heavy precipitation and subsequent flooding resulted in major power failures that affected the federal triangle area, where several agency headquarters and national cultural institutions are located (NCPC, 2008). Damage caused by the six-hour downpour on June 26 (considered a 200-year storm event) compromised building-monitoring security and high-speed communication systems, among other effects (Federal Triangle Stormwater Study Working Group, 2011).
"Snowmaggedon"	February 2010	A major snowstorm on February 5 through 6, 2010 dropped 20 inches of snow on the capital and left over 100,000 Potomac Electric Power Company (PEPCO) customers without power (Morrison et al., 2010). Called "Snowmaggedon," it was preceded by "Snowpocalypse" on December 19, 2009 (16–24 inches of snow) and followed closely by "Snoverkill" on February 9–10, 2010 (Samenow, 2011). February 2010 snowfall in the District totaled 32.1 inches (Mussoline, 2013).
North American derecho	June 2012	More than 107,000 PEPCO customers lost power due to strong thunderstorms and straight-line wind; some experienced blackouts for up to eight days (DDOE, 2012). Some DC residents were unable to reach 9-1-1 hotlines (FCC, 2013).
Heat wave	July 2012	A 1,000-foot section of Green Line track, one of the six subway lines servicing the District, had to be replaced due to heat-induced warping (Kunkle and Evans, 2012) after multiple days of temperatures exceeding 100°F.
Hurricane Sandy	October 2012	Twenty-five percent of cellular sites in affected areas (including the District) were disabled (Turetsky, 2013). More than 250,000 people in the Washington metro region lost power, but power was restored to 90% of customers within 48 hours (Preston et al., 2012).

D.1.1.3. *RISKS TO HUMAN HEALTH*

From a human health perspective, a study aggregating data on risk factors (age, poverty, linguistic isolation, educational attainment), land cover characteristics, and observed temperature patterns concluded that approximately 64% of the District's residents are at high risk of heat stress (Aubrecht and Özcelyn, 2013). Higher temperatures will increase the risk of vector-borne diseases (those transmitted to people from insects), such as West Nile virus. Health risks from urban heat also include heat stroke, dehydration, and respiratory diseases like asthma. The elderly, children, the ill, and the homeless are particularly vulnerable to these health risks (MWCOG, 2008). High rates of poverty and homelessness in DC make these health risks a particular concern. Approximately 17.8% of individuals and 14.5% of families in the District live in poverty (compared to 13.2% of individuals and 9.6% of families nationally; DDOT, 2010a). Studies have documented that among DC children, poverty is correlated with asthma (Babin et al., 2007). DC's homelessness rate is higher than that of any state and of all but four U.S. cities (Witte, 2012). Of the District's 4,300 homeless children, nearly one-fifth have asthma (Bassuk et al., 2011).

The District's ability to cope with extreme events (and more generally with climate change) depends on resources available at community and household levels. The greater Washington region is the fourth largest economy in the United States. It is also home to more *Inc. 5000* fastest-growing companies than any other U.S. city (WDCEP, 2010). DC is also home to the federal government, which accounts for approximately 34% of the city's employment and provides a measure of stability and access to resources. At the same time, the District's growth as a "strong and resilient economy" in the past decade is credited to its increased economic diversification, including the emergence of green businesses (Washington DC Economic Partnership, 2010).

In contrast to the robust resources available at the District and regional level, a great deal of vulnerability exists at the household level. As noted earlier, rates of poverty and homelessness in DC are above the national average. The unemployment rate as of March 2014 was 7.5%, relative to the national unemployment rate of 6.7% (BLS, 2015). Among residents 65 years and over, 18.2% live below the poverty line, 43.1% have no vehicle available, and 1.6% lack home telephone service (DDOT, 2010a). In 2010, one in five households in the District had a severe household burden (defined as housing costs that equal or exceed 50% of household income). In the very low-income bracket, that ratio was three in five (Reed, 2012).

D.1.1.4. *FLOODING AND IMPAIRED WATERS*

The District of Columbia lies within a region for which annual precipitation is projected to increase by anywhere from 4 to 27% over the next century (IPCC, 2007b). The temporal and geographic distribution of rainfall might change, and intense precipitation events might increase. Washington, DC is highly susceptible to flooding due to (1) its location between the Potomac and Anacostia Rivers near the entrance of the Potomac into the Chesapeake Bay; (2) the historic filling or burial of three streams in the DC area, which had been a natural drainage system; and (3) its low elevation and broad floodplains. Flood risks include overbank flooding of the Potomac or Anacostia Rivers, urban drainage flooding from undersized and combined sewers, and tidal/storm surge flooding (NCPC, 2008; Koster, 2011).

The National Mall Levee, part of the Potomac Park Levee System in downtown Washington, DC was built in 1936 to protect the city from flooding of the Potomac and Anacostia Rivers. The levee system currently has three open sections that must be closed during a flood event. The National Mall Levee, one part of the system, received an "unacceptable" rating from the U.S. Army Corps of Engineers (USACE) in 2007, leading to a de-accreditation by the Federal Emergency Management Agency (FEMA) and the release of new flood maps showing most of downtown DC without flood protection. To correct the issue, the USACE redesigned one closure (at 17th Street) and proposed making two closures (at 23rd Street and Fort McNair) permanent. These improvements reduced the District's chance of the levee being overtopped in any given year to less than 1% (NCPC, 2008). Work on the 17th Street levee was originally scheduled to be completed in 2011, but it was repeatedly delayed and finally completed in 2014 (USACE, 2014).

Approximately one-third of the District is served by combined storm and sanitary sewers that overflow into waterways if the flow exceeds the wastewater treatment plant's (WWTP's) capacity. The District is under a consent decree to build storage upstream of the WWTP to hold excess storm/wastewater in flood events and prevent overflow into waterways (NCPC, 2008). If the sewer main capacity is exceeded in extreme high-flow events, stormwater can back up into the streets. Also, if the sewer outfall is inundated by a high water level in the receiving stream, the sewers can back up. The DC Water and Sewer Authority (DCWASA) has installed gates at the outfall locations to help avoid these issues (NCPC, 2008).

Despite the projected increases in precipitation (which will replenish the Potomac River and nearby aquifers), climate change could adversely affect the District's drinking water supply. Higher temperatures might reduce the amount of precipitation that ultimately reaches the District's water sources. In addition, less precipitation falling as snow into the watershed and more falling as rain will lead to exaggerated seasonal runoff patterns (more streamflow in winter/spring and longer low-flow periods in summer), contributing to seasonal problems in water availability (Ahmed et al., 2013; MWCOG, 2008).

Climate change might also affect water quality and increase the burden placed on the water treatment facilities that serve DC (Ahmed et al., 2013; MWCOG, 2011a) by decreasing the raw water quality. For example, higher temperatures might contribute to increased algal blooms and lower oxygen levels. More intense precipitation could also lead to increased nonpoint source pollution (suspended sediment, nutrients, and chemical contaminants in rivers and lakes). Flooding could increase leaching from landfills, hazardous waste sites, and brownfield sites.

Threats to the District's landscape and built environment from more intense precipitation events include erosion; slope and roadway flooding and washout; roadway subsurface deterioration; tunnel flooding; road embankment failures; scouring of bridge and culvert abutments; culvert failures; drainage overloading and failure; tree and vegetation damage; power and other utility failure; increased occurrence of mold in buildings; stream degradation; effects on habitats and species; and changes in the water table that could affect development, septic systems, and the water supply (DDOT, 2013; MWCOG, 2011a, b, c, 2013a).

Tropical storms such as hurricanes are expected to be fewer in number but characterized by greater wind speeds and more intense precipitation (IPCC, 2007a).

D.1.1.5. *SEA LEVEL RISE*

Sea level rise threatens the District's military facilities, monuments, museums, federal agencies, roadways, bridges, metro lines, railroads, educational institutions, and fire stations. In DC, sea level has risen 3.16 millimeters per year on average since 1924 (a total of 0.3 meters or 15 inches; NOAA, 2013), and it is expected to rise further (Ayyub et al., 2012). Ayyub et al. (2012) modeled impacts of a 0.1, 0.4, 1, 2.5, and 5-meter sea level rise, which indicated that further sea level rise between 0.1 and 2.5 meters would inundate between 103 and 302 properties (residences, apartments, hotels, etc.) with combined property values between $2.1 billion and $6.1 billion (in 2005 dollars). A sea level rise of 5.0 meters would affect 1,225 properties with an assessed value of $24.6 billion.

Threats to the District's landscape and infrastructure from sea level rise include the loss of wetlands, erosion of roadway subsurface, bridge scouring, embankment failures, reduced vertical clearance for bridges, flooding of roadways in low-lying areas, changes in floodplains, and increased tunnel flooding (DDOT, 2013; MWCOG, 2011b, c, 2013a). Sea level rise may also increase the salinity of the coastal rivers that empty into the Chesapeake Bay. The salinity of the rivers will also increase during droughts and seasonal low-flow periods brought on by warming temperatures.

D.1.1.6. *ENERGY DISRUPTIONS*

As shown in Table 12, losing electricity is a common result of extreme weather events. Electricity comprises the majority of the District's energy infrastructure (70%), and it is more vulnerable to disruptions than the infrastructure for natural gas and petroleum (DDOE, 2012). The District Department of Environment (DDOE) cites two reasons for the greater vulnerability of electricity distribution. First, customers do not locally store electricity, so any disruption in electricity distribution is immediately felt by the end user. Second, distribution via overhead transmission lines makes electricity vulnerable to storm damage. Some 40% (approximately 101,200) of PEPCO's customers receive their electricity via aboveground power lines that are susceptible to fallen trees, heavy winds, and other hazards (DDOE, 2012; PEPCO, 2010). Only a very small fraction (0.1%) of energy consumed in the District is locally sourced (e.g., from solar), making the city vulnerable to disruptions in its external supply (DDOE, 2012).

Threats to the District's landscape and built environment associated with intense precipitation and flooding were noted earlier. Other storm-related or extreme event impacts (e.g., high winds) can cause damage to road surfaces, commuter/freight rail systems, bridges, and buildings; stress on the urban tree canopy; and power failures (MWCOG, 2011a, b, c, d, 2013a; DDOT, 2013). In addition, storms might disrupt other essential services (telecommunications, food distribution, water and wastewater services, etc.). Although DC has one of the most robust public transit systems in the country (MWCOG, 2008; Sustainable DC, 2013), the Mayor's Office has warned that the city's transportation infrastructure is growing old and becoming less resilient to extreme weather events (Sustainable DC, 2013).

D.1.2. REGION-WIDE ADAPTATION AND MITIGATION PLANNING

A city's resilience to climate change depends in part on the resources at its disposal and its economic strength. Measuring by GDP per metropolitan statistical area, the Washington–Arlington–Alexandria, DC–Virginia–Maryland–West Virginia metropolitan statistical area was the sixth largest economy in the United States in 2014 (BEA, 2015). When using wages, unemployment, growth rates, housing costs, and other variables to determine relative economic strength rather than size, DC usually remains highly ranked; for example, *Business Insider* ranked DC third in the nation for 2015 (Kiersz, 2015).

What makes planning and governance of the District unique among U.S. cities is the federal government's oversight authority. While DC is governed by its legislative body, the DC Council, the U.S. Congress oversees the DC Council, reviews the Council's actions, and can overturn some of the District's decisions and actions. Congressional oversight and the District's close coordination with federal agencies such as FEMA, EPA, and Department of Homeland Security are critical factors in the District's planning and implementation of adaptation measures.

In addition, like many U.S. cities, the District's adaptation planning has been influenced by the work of its regional council, which in this case is the Metropolitan Washington Council of Governments (MWCOG). A nonprofit organization, MWCOG is composed of 300 elected officials, representatives from 28 local jurisdictions in Virginia-Maryland-DC, along with Maryland and Virginia county and state government officials. Four council members participate on behalf of the District in MWCOG, and much of the District's current adaptation planning is based on MWCOG's work.

The Council is a resource-intensive organization with a significant role in coordinating data and research to undertake regional projects that would be difficult for one entity or local jurisdiction to accomplish alone, including those related to climate adaptation planning. MWCOG creates inter-municipal agreements for projects benefitting the region. It receives funding for studies and projects through various agreements between the local member jurisdictions and through federal grants. MWCOG coordinates a cooperative purchasing effort (across member municipalities), so the region benefits from economies and efficiencies of scale. Because it must compete for federal funding for its studies and projects, MWCOG ensures its projects closely match federal policies, objectives, and guidelines to keep regional efforts well coordinated in moving toward shared goals.

MWCOG prepares plans for regional (DC metro area) transportation, environment, housing, health and human services, homeland security, and public safety operations. The Council exerts a powerful regional influence. One of the best examples of a regional project planned and funded through the MWCOG is the Blue Plains Advanced Wastewater Treatment Plant that currently treats 43% of the metropolitan area's wastewater and should continue to do so for the next 40 years (MWCOG, 2013b).

MWCOG's Climate, Energy, and Environmental Policy Committee seeks to implement actions to respond to or lessen climate change-related impacts, including emissions mitigation. In 2008, the MWCOG Board adopted the National Capital Region Climate Change Report, which identified strategies to mitigate the effects of climate change, such as meeting greenhouse gas reduction targets (MWCOG, 2008). The report also included a range of adaptation strategies to address the eventual impacts of climate change. Adaptation planning identifies strategies and actions designed to decrease vulnerability to the immediate and long-term effects of climate change. The Committee manages and implements the measures in its 2008 report and the updated 2013–2016 Action Plan (MWCOG, 2013a). In planning these documents, MWCOG collaborated with stakeholders, EPA, and climate change experts. It released summaries of potential climate change impacts, vulnerabilities, and adaptation strategies for the region that will be in a future guidebook (MWCOG, 2013c). MWCOG plans and measures performance using data on climate-related drivers expected to affect the Mid-Atlantic region, which it categorizes as urban island heat, variations in precipitation, severe storms, and sea level rise over the next 50 years.

MWCOG forecasts that by 2030 the DC metro area will gain 1.6 million new residents and 1.2 million new jobs. MWCOG has estimated that greenhouse gas emissions will increase by 33% by 2030 and 43% by 2050. Two MWCOG reports set forth regional plans for water and air, healthy neighborhoods, resilient economies, and access to alternative housing and transportation. Goals, targets, and measurements of progress appear in four broad categories: accessibility, sustainability, prosperity, and livability (MWCOG, 2008, 2010). Similar to the Sustainable DC Plan described above, the MWCOG 2013–2016 Action Plan sets goals through 2020 for the District-Virginia-Maryland region to study, measure, and implement actions concerning the built environment and infrastructure, regional greenhouse gas emissions, renewable energy, transportation and land use, sustainability, and resiliency and outreach (MWCOG, 2013a).

Additionally, important work relevant to climate resilience was undertaken in the aftermath of the extreme events listed in Table 12. For example, a Federal Triangle Stormwater Study Working Group (2011) convened after the June 2006 downpour and flash flood noted how facility managers and service providers developed strong working relationships in the wake of the event, an experience noted also after Hurricane Irene in 2011. They continue to share short- and long-term flood-proofing strategies (Federal Triangle Stormwater Study Working Group, 2011).

D.1.3. CITY-WIDE PLANNING

In the District, climate change adaptation planning occurs in several departments, including the Mayor's Office, DDOT, DDOE, DCWASA, and the DC Homeland Security and Emergency Management Agency (DC HSEMA).[2] To date, these entities have developed the following plans and are implementing the recommended measures:

- Sustainable DC Plan (Sustainable DC, 2012; 2015)
- *DDOT Climate Adaptation Plan* (DDOT, 2013)
- DDOT *Action Agenda—Progress Report 2010* (DDOT, 2010b)

---

[2] More than half a million people live in Washington, DC, and the District's government includes more than 40 agencies or departments (2013, Mayor's office at dc.gov). Many other departments not mentioned in this report also contribute data and personnel to the District's adaptation planning.

- DDOE *Climate of Opportunity: A Climate Action Plan for the District of Columbia* (DDOE, 2011)
- DCWASA *Long-Term Control Plan Modification for Green Infrastructure* (DCWASA, 2014)
- DC HSEMA *District Response Plan* (DC HSEMA, 2008)

In addition, DC HSEMA is in the process of collecting feedback from the District Preparedness System's (DPS) public and private partners as part of an effort to develop a comprehensive District Hazard and Vulnerability Analysis. These plans cover the economy, energy, water, land use/land cover, the natural environment, people, telecommunications, and transportation. Together, the plans and action measures, along with regional efforts (discussed later on), form the basis of the District's current broad climate adaptation planning. It should be noted that because of DC's unique role as the nation's capital, a certain amount of redundancy between the responsible parties and actions has been purposely built into all of the District's adaptation planning.

The District Department of Health has also partnered with the RAND Corporation on Resilient DC, a program to build community preparedness and resilience (RAND, 2013). The focus of the effort is on building partnerships and collaborations among organizations in communities to leverage existing expertise and capacity, as well as reach out to underserved and vulnerable subpopulations.

All of the departments in the above list followed a rigorous process in developing their plans. For example, to prepare the 2013 Sustainable DC Plan, the Mayor's office and DDOE held public meetings with two key advisory groups: the Green Ribbon Committee encompassing the public, private and nonprofit sectors, and the Green Cabinet composed of DC agency directors. One goal of those meetings was to promote interagency coordination on the shared and individual agency missions and actions as they relate to the overall plan. The public meetings and discussions involved more than 4,700 people and allowed all involved departments to solicit feedback and opinions from members of the general public. After the public meetings, nine working groups of experts, DC government officials, and members of the public were created to address energy, food, climate, the built environment, nature, transportation, water and waste, and the green economy. The resulting Sustainable DC Plan is a citywide initiative to deal with a changing climate. Its overarching goals are to create jobs and economic growth, improve health and well-being, increase equity and opportunity, and preserve and protect the environment. The plan covers both climate change mitigation and adaptation.

D.1.4. CITY-WIDE ADAPTATION MEASURES

The sections below present the current goals and measures the District is carrying out in eight broad areas: climate/environment, built environment, energy, food, water, stormwater/wastewater, transportation, and nature/green space/trees. These measures are from the Sustainable DC Plan, except where noted.

D.1.4.1. *CLIMATE/ENVIRONMENT*

The District intends to advance physical adaptation and human preparedness to increase resilience to future climate change through the Sustainable DC Plan's climate goals. By 2032, DC will:

- Require climate change impact analyses as part of all new DC construction projects
- Assess its energy infrastructure's vulnerability to climate change, given that past power outages resulted from severe weather events
- Have DC emergency services, utilities, and disaster preparedness agencies respond more quickly and efficiently to climate-related weather emergencies
- Require new housing developments to integrate climate adaptation solutions into cost-effective building strategies, so that buildings last for 50 years or more

D.1.4.2. *BUILT ENVIRONMENT*

The Sustainable DC Plan tackles building codes and construction planning by setting a goal of net-zero energy use for all new construction projects by 2032. Specifically, the District will:

- Update its Green Building Act of 2006 and its Leadership in Energy and Environmental Design (LEED®) certification standards for facilities that are 50,000 square feet or larger
- Provide incentives for LEED Gold standard certification to ensure that future buildings will be resilient to climate change
- Require neighborhood-scale sustainability goals for all major redevelopment projects (e.g., Walter Reed Army Medical Center)
- Adopt the 2012 International Green Construction Code or an equivalent for all new construction and major renovations

D.1.4.3. *ENERGY*

By 2032, the District intends to reduce power outages to less than 100 minutes per year through energy infrastructure improvements. DC officials will work with stakeholders to add local renewable energy sources and decentralize its energy sources into a more effective power grid.

Starting in 2014, the District began a multiyear, $1-billion project to move high-voltage feeder lines underground, spearheaded by the District of Columbia Power Line Undergrounding Task Force (DCOCA, 2014) in order to reduce this vulnerability. DDOT and PEPCO jointly implement this public–private project.

D.1.4.4. *FOOD*

Because increased local food production can improve the District's resilience to climate change, DC intends to boost its agricultural land use by 20 acres by 2032. Specific measures include the following actions:

- Adopt the Sustainable Urban Agriculture Act and set up urban greenhouses and agriculture projects, in particular beekeeping
- Evaluate the potential for rooftop gardens and use of public parks and recreation areas for growing plots to streamline the process of finding land for community agriculture
- Retrofit at least 50% of DC public schools with gardens and integrate the planning, planting, tending, and harvesting of those gardens into the curriculum
- Make temporary agricultural sites for gardens available wherever possible

The Plan recognizes that the role the food sector plays in the DC economy can be increased. With that goal in mind, the District intends to produce or obtain 25% of its food within a 100-mile radius. Specific measures include the following:

- Initiate a comprehensive study on the sources of the District's food supply, ways in which that supply can become more localized, and sales of food from community gardens.
- Set up a nonprofit Food Policy Council to research the local food sector with the goal of providing nutritious food through a self-sustaining system.
- Purchase locally grown food for the DC public schools and government events.

D.1.4.5. *WATER—WETLANDS*

The Sustainable DC Plan intends to help residents and businesses adapt to climate change. It aims to protect the District against future flood risks by restoring wetlands and creating green infrastructure for stormwater drainage. Expanded green areas will help mitigate rising temperatures. Additional tree canopy will benefit the environment and District residents. The following actions aim to preserve and enhance wetlands, and thus have a climate adaptation dimension:

- Increase the wetlands along the Anacostia and Potomac Rivers by 140 acres or an additional 50% by 2032.
- Coordinate open space guidelines with the National Park Service to control invasive species.
- Develop an Urban Wetland Registry to be created by DDOE's wetlands conservation planning team.

- Restore habitat and biodiversity of the rivers through the Urban Wetland Registry.
- Require low-impact development planning for new waterfront development greater than 50,000 square feet, along with wetlands preservation activities.

D.1.4.6. *WATER—STORMWATER/WASTEWATER*

To reduce flooding and improve stormwater infrastructure by 2032, the District plans to:

- Use or capture 75% of its stormwater
- Install 2 million square feet of green roofs with the help of a rebate program
- Build an additional 2 million square feet of planted surfaces on public and private buildings by 2018[3]
- Add extensive green infrastructure elements for paved surfaces to capture pollutants and reduce runoff
- Double the number of homes in the DC RiverSmart Homes program for preventing runoff by using green technologies
- Replace gravel and impervious surfaces in alleys with permeable surfaces to create 25 miles of green alleys
- Institute new or revised zoning requirements for housing developments to improve stormwater retention
- Revise building codes to allow alternative water collection systems
- Increase the use of green infrastructure in public right-of-ways
- Provide financial incentives to promote efficient water use for landscaping and building
- Promote water conservation through improved metering and monitoring for leaks, etc., with alert systems

Outside of the Sustainable DC Plan, the District has also made several other water-related adaptation planning efforts, many in response to the 2005 Consent Decree from EPA and the Department of Justice (DOJ) that required the District of Columbia Water and Sewer Authority (DC Water) to design and construct underground storage tunnels to hold contaminated wastewater during storms and wet weather, with the goal of reducing combined sewer overflow (CSO) discharges. The largest of these is the DCWASA's DC Clean River Project, a 20-year, multibillion-dollar ongoing project consistent with EPA's policy directives for adaptive management and in line with the requirements of the 2005 Consent Decree (DCWASA, 2012). The Decree also required DC Water to promote green infrastructure as another approach to CSO control. The Clean River Project includes demonstration projects, public involvement, and green

---

[3] With more than 2.5 million square feet of green roofs, the District ranks highest among North American cities (GRHC, 2013). Green roofs and urban tree canopies contribute to community resilience by improving air and water quality, moderating the urban heat island effect, reducing energy consumption, providing recreational opportunities, mitigating flood impacts, and providing ecosystem services (Rodbell and Marshall, 2009; GRHC, 2013).

infrastructure improvements in construction and land use, such as bioswales, green roofs, permeable pavement, and other green technologies. Further, DC expects to reduce 96% of its combined sewer overflows by using inflatable dams, rehabilitating pump stations, and adding separate municipal storm sewer systems. Throughout the 20-year project, the District will focus on meeting the requirements of the 2005 Consent Decree and EPA water quality standards. The Potomac River, Rock Creek, and Anacostia Rivers are the focus of the planned improvements, and the project requires coordination with DDOT for easements, such as the Blue Plains Tunnel and other infrastructure improvements. The District hopes to benefit from the state-of-the-art implementation, which should make the District's wastewater system more resilient to extreme weather and precipitation events (DCWASA, 2012).

On May 19, 2015, the First Amendment to the 2005 Consent Decree was lodged and opened for public comment. The Amendment requires DC Water to implement green infrastructure as part of the existing DC Clean Rivers Project. In anticipation of the amendment's approval and in agreement with EPA, DOJ, and the District, DC Water announced a Green Infrastructure Plan in 2015 (DCWASA, 2015). The plan modifies the existing DC Clean Rivers Project, a $2.6-billion project to limit untreated sewage flow into area rivers through the construction of new tunnels. Under the Green Infrastructure Plan, some proposed tunnels will not be built; instead, the stormwater capacity intended for the tunnels will be mitigated through green infrastructure, such as infiltration basins and green roofs. This will allow for rainfall to infiltrate soils before it becomes stormwater runoff, alleviating the need for costly and disruptive tunnels. The new approach allows for faster implementation and potentially boosts property values near restored natural areas.

In 2013, the DDOE released new stormwater management regulations, which require new development or substantial redevelopment to meet standards for onsite water retention (DDOE, 2013a). The goals of the regulations are to increase infiltration and decrease runoff to protect area waterways and comply with federal clean water standards, as well as to create a more equitable distribution of stormwater throughout the District. This contrasts the former regulations, which focused on the timing and quality of stormwater throughout the District, not reducing the overall quantity of stormwater generated. To create a financial incentive for change, voluntary retrofits accumulate stormwater retention credits, which can be bought by developers to offset required reductions at other sites. The District further attempted to reduce stormwater by adopting new construction codes based on the 2012 International Green Construction Code. These code changes support increased onsite use of rainwater to reduce stormwater generation.

The District has also developed more focused adaptation planning. In 2012, the Mayor's Task Force on the Prevention of Flooding in Bloomingdale and LeDroit Park, two DC neighborhoods, issued final short-, medium-, and long-term recommendations to reduce the chance of severe flooding. These neighborhoods are serviced by inadequate late nineteenth-century combined sewer and stormwater infrastructure, which has resulted in floods of mixed raw sewage and stormwater during intense precipitation events, posing numerous health and safety risks to residents, rescuers, and repair crews. As a long-term solution, the DC Clean Rivers Project will include building an estimated 600-million-dollar tunnel system 5 miles in length to provide excess capacity (DCOCA, 2012).

D.1.4.7. *TRANSPORTATION*

The transportation plans and measures adopted by the DDOT further strengthen the Sustainable DC Plan's goal of making the District's transportation infrastructure capable of withstanding the upper limits of projected climate change impacts by supporting DDOT in its use of climate change indicator data. DDOT uses three planning documents as the bases for improving the District's resilience to climate change: (1) the *Climate Change Adaptation Plan* (DDOT, 2013), (2) the DDOT *Action Agenda—Progress Report 2010* (DDOT, 2010b), and (3) the DDOT Urban Forestry Administration (UFA)*'s District of Columbia Assessment of Urban Forest Resources and Strategy* (DDOT, 2010c).

The 2013 *Climate Change Adaptation Plan* includes the District's vulnerability assessment for transportation infrastructure and the corresponding adaptation planning and measures to promote resilience to extremes in temperature, precipitation, sea level rise, and storms for its 4,000 miles of roads, 240 bridges and tunnels, and watershed with associated trees and vegetation. In planning and decision-making efforts, DDOT used the National Cooperative Highway Research Program assessment tool to define the scope of its needs, access vulnerability, and integrate the information collected. DDOT chose indicators, such as sea level rise or temperature, for each category of community assets (e.g., bridges, trees), listed impacts, and ranked vulnerabilities as high, medium, or low for each indicator. Potential adaptation strategies that DDOT plans to use include DDOT climate projection models through 2100; vulnerability assessments; staff training; updating design standards and policies; updating potential strategies for adoption and use in all new projects; coordinating with other agencies; and seeking funding for assets (DDOT, 2013).

The 2010 *Action Agenda—Progress Report* highlights DDOT's low carbon footprint initiatives, including establishing bike lanes throughout the District, and a bike share program with 100 stations and 1,000 bicycles. DDOT is taking action to promote walking, bus-riding, and greater use of the metro system to meet the challenges of this century and the next. To reduce stormwater runoff, urban heat, and energy use, DDOT plans to retain all stormwater from rainstorms of at least a 1.2-inches, use 15% less energy, provide electric car recharge stations, use low-impact development, and provide public outreach on various adaptation measures. DDOT already installed 1,200 solar-powered parking meters (saving energy) and interactive electronic devices in bus shelters to provide real-time bus information to the public (as part of outreach). Bus, metro, bike share, parking meters, and other pay-per-use transportation features will operate on a "one card" system for all (DDOT, 2010b).

Additionally, the Washington Metropolitan Area Transit Authority received $20 million in Hurricane Sandy disaster recovery funds to invest in flood mitigation for MetroRail (U.S. DOT, 2014). The majority of the money was spent upgrading venting structures to prevent flood water from entering the system, while the remainder was spent on drainage improvements.

D.1.4.8. *NATURE/GREEN SPACE/TREES*

A recent assessment found that tree canopy covers 35% of all land in the District (DDOT, 2010c), compared with an average tree cover of 30% measured across 18 major U.S. cities[4] (Nowak and Greenfield, 2012). By 2032, the District intends to cover 40% of its land with tree canopy (DDOE, 2013b) by planting 8,600 new trees per year through 2032 using heat-tolerant species that will be more resilient to climate change. The District already has, according to the plan, imposed a Green Area Ratio requirement for land use in all new development sites to improve stormwater management, air quality, and urban heat island effects.

DDOT's UFA currently manages approximately 144,000 trees on streets, in parks, and in recreation areas. DDOT believes that trees are one of the District's most important assets. The leaves help shade people and buildings during heat waves, and the roots help trap water and soil in place. Urban trees also prevent runoff, absorb pollutants, and reduce urban heat island effects. The UFA's 2010 *Assessment of Urban Forest Resources and Strategy* is a plan to increase the urban canopy; protect and improve air and water quality; and build capacity in its community forest program. Between 2006 and 2011, the District increased its tree canopy by 2.1 to 37.2% (DDOT, 2011). The strategy includes actions that will promote resilience of the natural environment, as well as water and air quality. Recent weather events, such as Hurricane Sandy and the 2012 derecho, caused significant tree loss and damage.

D.1.5. CITY-WIDE EMERGENCY RESPONSE

Although climate change planning and adaptation are not the express purpose of the DC HSEMA Response Plan, the plan does cover extreme weather events. It covers traditional response elements and involves multiple and redundant agencies, systems, and measures to increase responsiveness and resilience. The communication, coordination, and control systems between the Mayor's Office and local and federal agencies are thoroughly delineated in the plan.

Through DC HSEMA, the District addresses long-term disaster planning (as well as strategic planning) that includes permanently replacing housing, dealing with environmental pollution, and restoring infrastructure. The District Response Plan also addresses services for vulnerable populations and the general public and builds in redundant public health and emergency response systems. Because the plan includes so many different DC and federal agencies and because the area has recently experienced a wide range of extreme events, the plan is used, tested, and updated often. Staff, funding, and equipment are available within close proximity for almost any emergency situation in the District.

Discussions with public health professionals in the DC metropolitan area determined that, although no one agency is legally charged with coordination in an emergency, informal relationships are well established among local and state health departments and other public health partners, resulting in strong regional coordination (Stoto and Morse, 2008).

---

[4] The 18 major U.S. cities examined in this study were Albuquerque, NM; Atlanta, GA; Baltimore, MD; Boston, MA; Chicago, IL; Denver, CO; Houston, TX; Kansas City, MO; Los Angeles, CA; Miami, FL; Minneapolis, MN; Nashville, TN; New York, NY; Pittsburgh, PA; Portland, OR; Spokane, WA; and Tacoma, WA.

D.1.6. DATA COLLECTION APPROACH

In the case of Washington, DC, the project team convened participants from across the District government for an initial and a follow-up workshop. Throughout this process, the project team worked closely with the DDOE to identify participants, understand previous or planned resilience and adaptation efforts in the District, and hold the workshops. Both workshops included sessions in which participants provided data for the quantitative indicators and scored the qualitative indicators. The project team's presentations at the beginning of each workshop introduced participants to the tool's methodology and goals. The workshops also included presentations by DDOE and the project team on existing resilience and adaptation work in DC. A list of workshop attendees is provided in Appendix B. Full agendas for the workshops are provided in Appendix C.

DDOE identified workshop participants who manage activities within some of the eight sectors identified in the tool from agencies across the government. DDOE also identified workshop participants who operate public services (e.g., public transportation). Most of these participants had previously joined in DDOE-led sustainability or resilience efforts. Each sector had at least one participant with in-depth knowledge of operations and status in that sector. Because the project did not intend to achieve consensus or to quantify differences among participants, each sector had one individual or a group of two to three individuals designated as the expert (or experts) charged with tool implementation activities. Some sectors had more than one individual in this role because they covered a broad range of topics (e.g., the water sector required experts on drinking water quality, drinking water supply, and wastewater).

The initial workshop began with a presentation by the project team introducing the project and a presentation by DDOE that provided additional overviews of previous or ongoing climate resilience work in the District. The project team presented on the overall tool methodology, including details on how to use the question component. At the time of the first DC workshop, the thresholds were not yet developed. To provide the participants at this workshop with a baseline resilience score, a climate change resilience expert at DDOE drafted a resilience score for indicators of which he had knowledge. Participants then divided into breakout groups, determined by the project team, for each sector. Each breakout group was provided with a facilitator trained on the tool, a note-taker to capture the discussions, and printed handouts of the questions for the qualitative indicators. With this support, the breakout groups provided importance weights and resilience scores for the questions pertaining to their sector. Following this session, the workshop continued with a presentation by the project team and DDOE on climate adaptation work in the District and an overview by the project team on the tool's indicator component. Breakout groups then reconvened to provide importance weights and resilience scores for the indicators and suggest any relevant data sources that the team had not previously identified. The workshop concluded with a debriefing that asked for participant feedback on the tool and the process.

After the first workshop, the project team analyzed the results and communicated with some participants individually to obtain clarification on results or suggested data sources. The project team then convened a follow-up workshop to present additional data identified during the first workshop, gather additional information, and provide clarification on some qualitative and quantitative indicators. Thresholds had also been developed for the tool for use at any site. This methodology provided participants with guidance on resilience scoring throughout the process. Due to individuals' availability, the group of participants at the follow-up workshop was slightly different from the initial group. Thus, the follow-up workshop also began with a presentation to review the project and the tool methodology. The first part of the workshop also included a presentation by DDOE on the District's progress on developing a climate adaptation plan. The project team presented preliminary results from the initial workshop. In the breakout sessions that followed, participants:

- Reviewed scoring for qualitative and quantitative indicators that the project team had modified based on suggestions from the first workshop
- Selected the most appropriate data set for quantitative indicators for which participants at the initial workshop had suggested alternate data sets
- Provided any additional data or data suggestions

The follow-up workshop ended with a debriefing session during which the project team asked participants to consider which sectors might contribute most to climate resilience in the District. The project team also asked participants to suggest ways of displaying results that would most benefit continued resilience and adaptation work in the District. Appendix D includes the graphical representations of the results from the two workshops.

D.2. WASHINGTON, DC Results

D.2.1. CITY-WIDE RESULTS

The average results on resilience and importance across all sectors in Washington, DC, based on participants' responses to questions as qualitative indicators and the importance weight assigned to each, are summarized in Figure 3. The same information is supplied for indicators in Figure 4.

For both resilience and importance, scores ranged from 1 to 4, with one indicating lowest resilience or lowest importance, and 4 indicating highest resilience or highest importance. In Figures 3 and 4, the "resilience score" represents an average score for all qualitative or quantitative indicators in that sector. These sectors are ranked, from left to right, by the average importance score for that sector. As such, a sector with a low resilience score towards the right of the plot may be considered relatively vulnerable compared to another sector with a low resilience score towards the left.

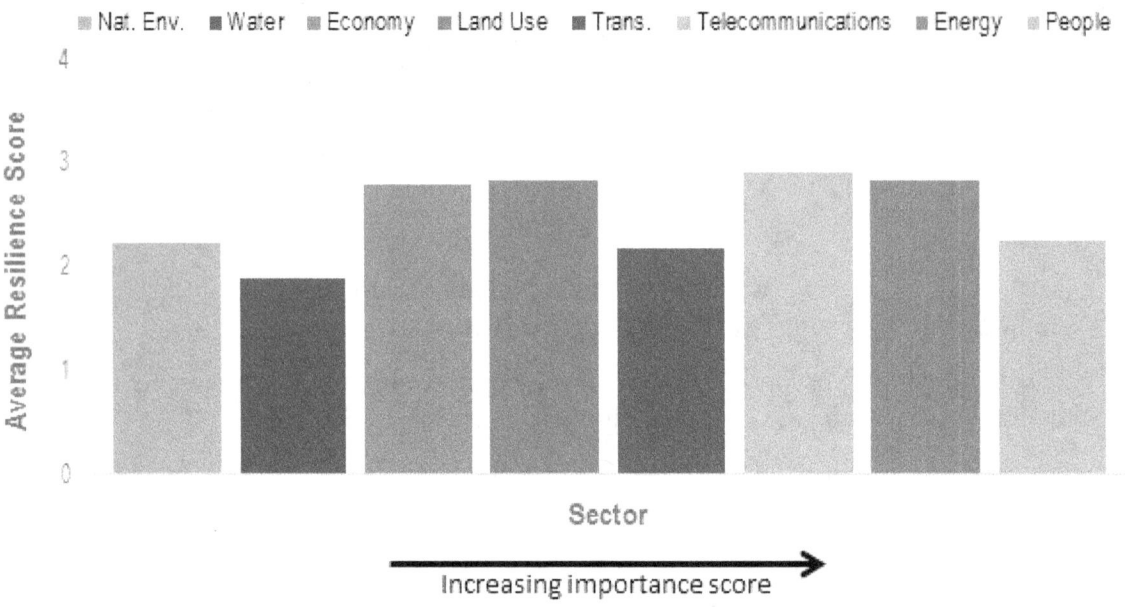

Figure 3. Washington, DC: Average qualitative indicator resilience and importance.

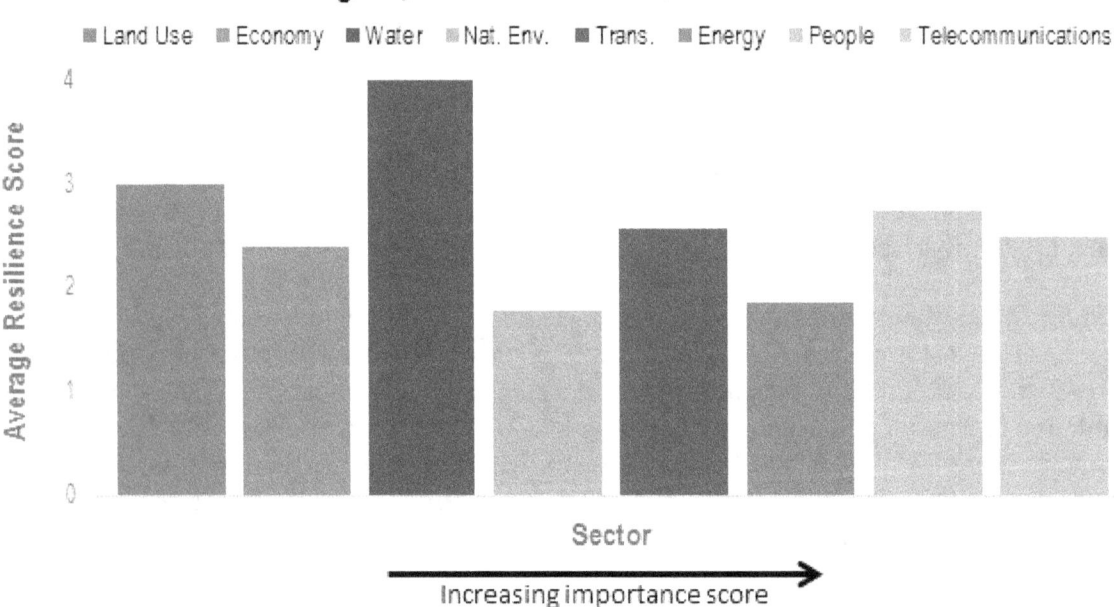

Figure 4. Washington, DC: Average quantitative indicator resilience and importance.

For the qualitative indicators, no sectors received an average score of greater than 3 or less than 1. For the importance scores, results were similarly clustered, although the overall scores were higher. On average, there were no sectors that scored in the top quartile of importance but the lowest quartile for resilience—a situation that would suggest high vulnerability across the entire sector. The people, transportation, and water sectors had lower resilience scores but similar importance scores to the other sectors, suggesting that these sectors in Washington, DC may be more vulnerable to the effects of climate change, and impacts to these sectors will create more significant disruptions.

For the quantitative indicators, there is much more variability in scores across sectors than in the qualitative indicator data. The greater variability in the quantitative indicator data may be due to limitations in the available data sets that focus on a particular subset or area of the sector that may be performing better or worse than the sector overall. With the qualitative indicators, the project team had fewer obstacles to achieving a comprehensive picture of resilience across all issues that might affect the resilience of a sector. However, the quantitative indicators still add value to the overall analysis.

Figures 3 and 4 convey differing narratives for citywide preparedness. While Figure 3 suggests that in Washington, DC, no one sector is more in need of urgent attention (high importance and low resilience), Figure 4 highlights that, based on data available, the natural environment and energy sectors both have lower resilience scores and similar importance scores compared to other sectors, suggesting that these two sectors may need more attention. By contrast, the water sector has high average resilience and relatively low average importance, so it may not be as critical to focus on this sector.

Additionally, there may be more localized risks within and across sectors. Therefore, while the averages presented in in Figures 3 and 4 help identify an overall trend, they may also mask important data points, increasing the risk of concluding that there is no evidence for action when action is warranted (i.e., type II errors).

Figures 5 and 6 disaggregate the data summarized in Figures 3 and 4, and they highlight potential "spikes" of high risk within sectors with overall lower averages. Both Figures 5 and 6 confirm the potential for type II error because many of the sectors show significant spread across both the resilience and importance score axes.

Figures 5 and 6 also indicate the possible action pathways stemming from the results, and they show that the District faces a significant number of moderate to highly critical vulnerabilities to address across all eight sectors, along with a potential need for increased monitoring. This is true for both the question (see Figure 5) and indicator (see Figure 6) data. Comparatively, there are few low-priority items and small problems. Overall, as indicated in Figures 3 and 4, the water, transportation, and people sectors appear to pose the greatest concerns in terms of resilience.

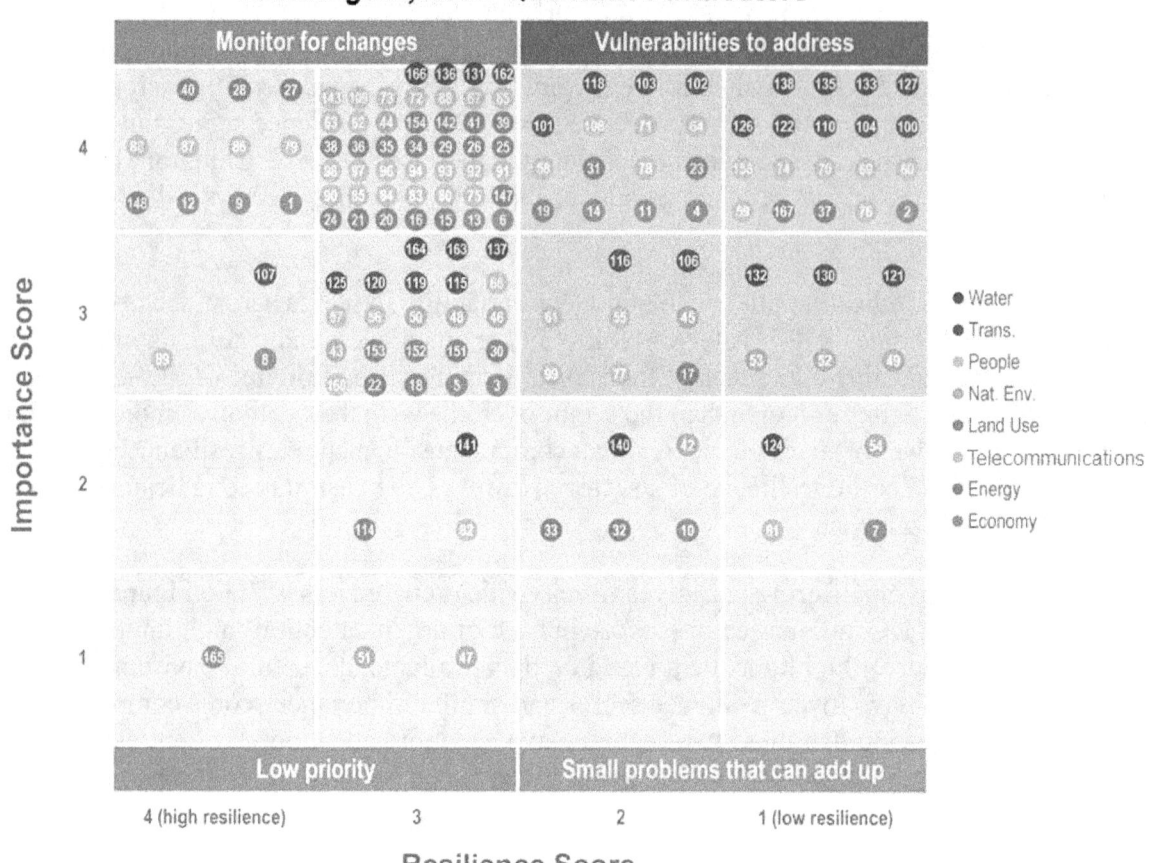

Figure 5. Washington, DC: Qualitative indicator quadrant mapping.

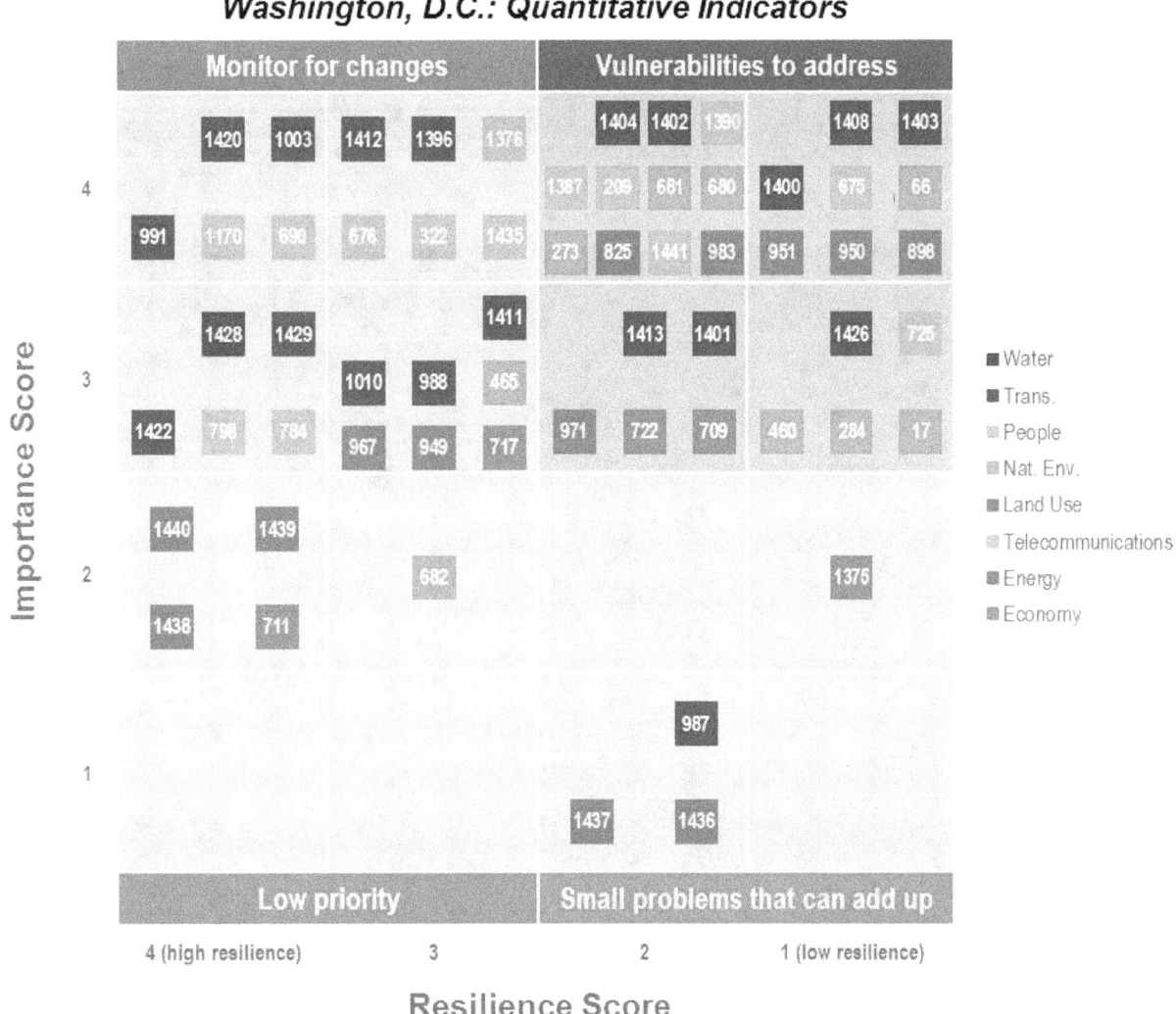

Figure 6. Washington, DC: Quantitative indicator quadrant mapping.

D.2.2. SECTOR-SPECIFIC INVESTIGATIONS

The sector-specific discussions below connect the results of the workshop exercises to potential underlying drivers and roadblocks for each sector discussed in the existing literature. Workshop participants also provided additional insight into each sector when giving additional information regarding the assigned importance and resilience scores.

D.2.2.1. ECONOMY

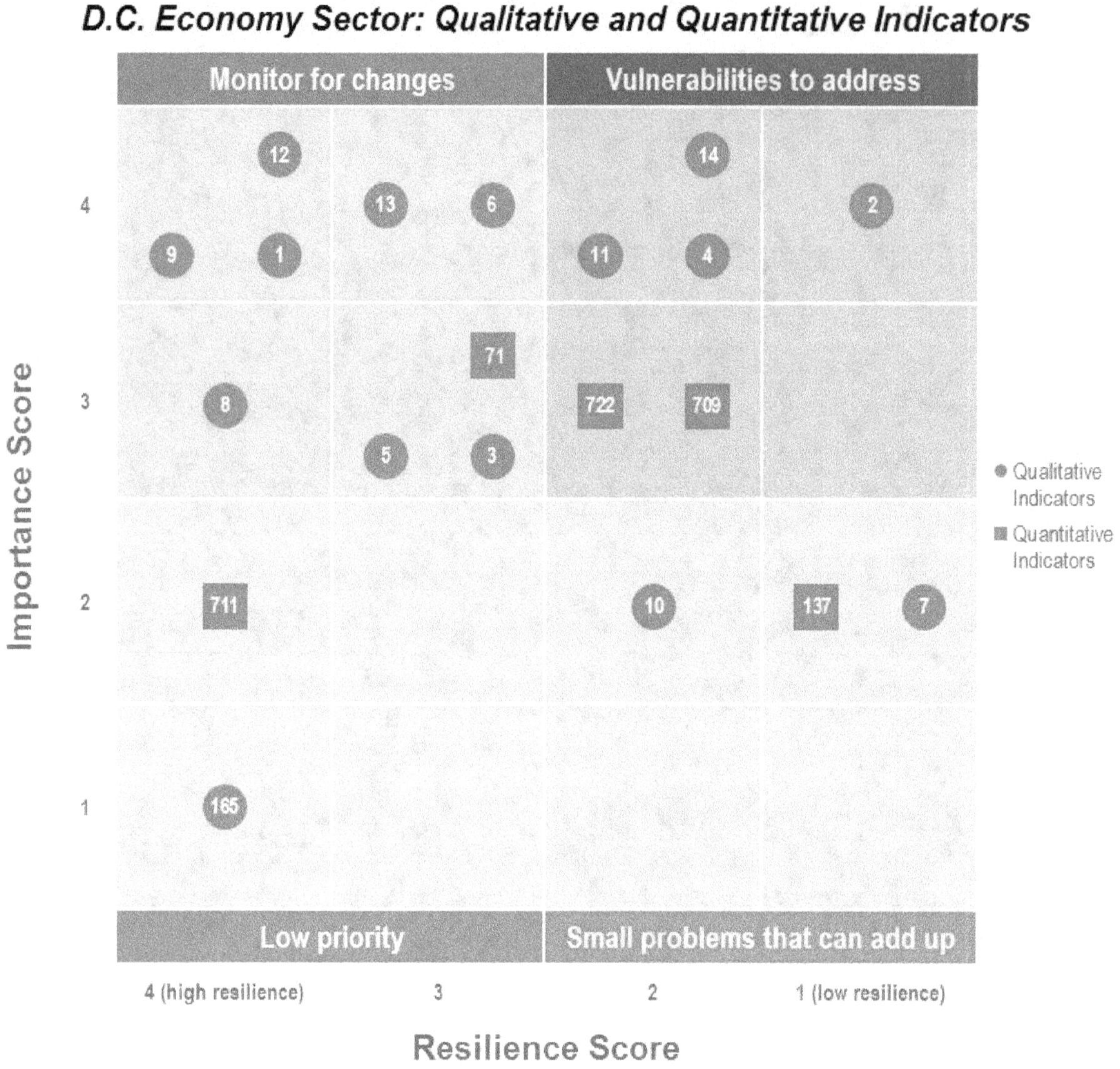

Figure 7. Washington, DC economy sector: Qualitative and quantitative indicator quadrant mapping.

Overall, workshop participants and supporting data indicate that Washington, DC's economy sector is independent, diverse, and robust. Washington, DC is an economic center and operates independently of Maryland and Virginia. DC employment centers are also very diverse, which underlies the District economy's resilience to climate change. The District has also taken steps to understand the potential impact of climate-related events on the local economy (e.g., the impact of major changes in energy policy).

The District leverages current resources to perform effective adaption planning and increase resilience. According to workshop participants, adaption planning successfully considers costs and benefits, encourages pre- or post-event effectiveness evaluations, and frequently involves analyses of past climate-related events. Furthermore, existing disaster response planning increases resilience. The District Response Plan identifies which agency is responsible for each function within disaster response needs. For example, during Hurricane Sandy, the District government closed, but the agencies needed for communication or support were activated and in emergency support centers. Workshop participants noted, however, that the planning process is only somewhat flexible and that no mechanisms are in place to help businesses return to normal operations after an extreme weather event. Additionally, adaption plans account for few resilience–cost tradeoffs between the less resilient but lower-cost strategy of increasing protection from climatic changes, and the more resilient but higher-cost strategy of moving residents from the most vulnerable portions of the urban area; this critical factor is discussed below in relation to intracity disparities. While the District has successful adaptation planning processes, the lack of flexibility in the planning process and few considerations of resilience–cost tradeoffs can reduce the its effectiveness, thus decreasing the economy's resilience.

However, resilience scoring based on economic indicator data was mixed. The results indicate that while the District's economy may appear to be relatively resilient based on a District-wide indicator, there may in fact be significant intracity disparities. For example, in 2012, the District's unemployment rate was relatively moderate (8.9%) and in 2011, 92.9% of the noninstitutionalized population had health insurance, indicating high economic resilience. However, these data mask a significant range in values across the District; in 2012, one ward had a 2.8% unemployment rate, while another had a 22.4% unemployment rate.

In addition, approximately 18.2% of persons in the District live below the poverty line, indicating low resilience. Again, however, this indicator does not reflect intracity disparity. High-poverty areas tend to be in low-lying areas, which are more vulnerable to sea level rise, storms, and other extreme weather events resulting from climate change. However, workshop participants did not rate this indicator as particularly important in the economy sector. A higher importance score was given to the percentage of owned housing units that are affordable (33.7%). Workshop participants noted that DC has many vulnerable people with a high housing burden.

Finally, indicator results may also mask disparities related to timing rather than geography. For example, the District experienced a 1.27% decrease in its homeless population from 2012 to 2013, indicating moderate resilience. However, workshop participants noted that DC might be less resilient than the data suggest because DC has instituted an absolute right to shelter during hypothermia season, so the point-in-time count of homeless persons in June is very different than in January.

Figure 7 shows that 45% of the qualitative and quantitative indicators lie in the "monitor for changes" quadrant (high resilience/high importance). In addition, most qualitative and quantitative indicators (75%) are above the median for importance. These trends indicate that the District has begun to recognize and address the need to have a resilient economy in the face of climate change. There is room for improvement to ensure the District's economic resilience to climate change, as 30% of the qualitative and quantitative indicators fall in the "vulnerabilities to address" quadrant (low resilience/high importance).

D.2.2.2. ENERGY

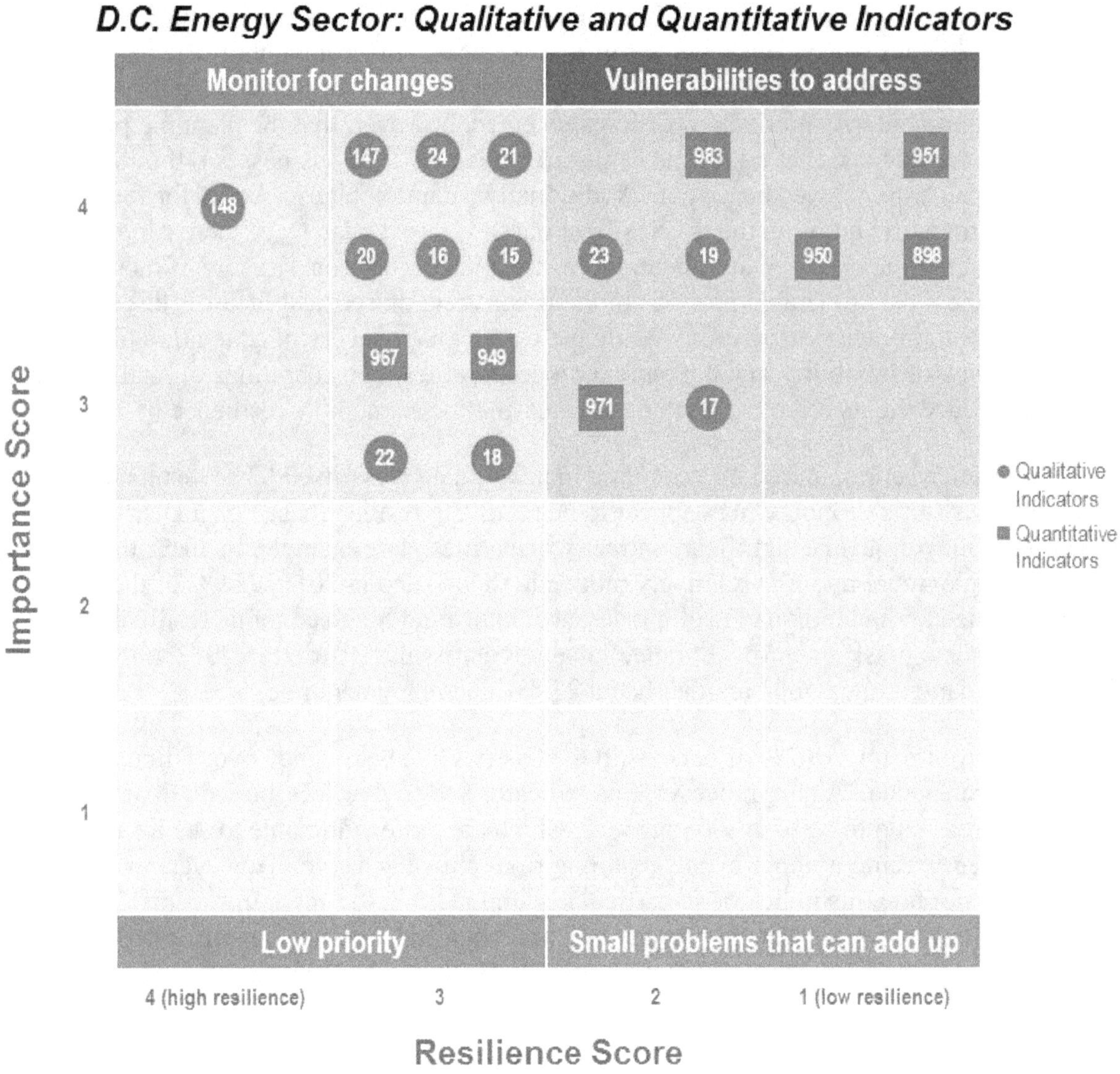

Figure 8. Washington, DC, energy sector: Qualitative and quantitative indicator quadrant mapping.

Washington, DC is generally resilient with respect to energy supply. The District has a diverse energy portfolio and redundant systems are in place for coping with extreme events at the regional level, although coverage may be inadequate at the customer or building level. The total energy source capacity per capita is 4.2 kilowatts, which indicates high resilience. In 2010, electricity accounted for the majority of energy consumed in the District at 70.4%, followed by natural gas (18.3%), petroleum (11.3%), and renewable sources (a low 2%). The District's main electricity provider, PEPCO, runs a peak energy savings program that encourages customers to track their energy use and incentivizes peak use reduction. Peaking plants in Maryland and Virginia can help the system cope with higher peak demands at different times than currently experienced. PEPCO has developed plans to address potential increases in electricity for cooling.

Although energy supply is generally resilient, several factors make other areas of the energy sector less resilient. Most of the energy supply originates outside of the District, in Maryland, Virginia, and West Virginia, and the diversified generation of energy does not currently occur in the District. According to workshop participants, the political and technical capacity could allow generation from multiple sources. The District also reported high energy use per capita. In 2010, average electricity use per capita in the District was estimated as 15,034 kilowatt-hours, above the national average of 12,954 kilowatt-hours and the third highest national per capita use (World Bank, 2015). Another source (U.S. EIA, 2013) reported average total energy use at 208 million British thermal units per capita. The capacity of the District's source per service area is also low at 13.28 million gallons per square mile. These values indicate overall low resilience to climate change, although the District does have available smart grid opportunities to manage demand.

In terms of power outages, the resilience of the District is mixed. Based on average power outages per year, the District has low resilience to climate change. However, workshop participants disagreed with indicator thresholds, noting that the range of 1 to 24 hours associated with a resilience score of 2 is too large, as residents can generally tolerate 1 to 2 hours without power. A full 24 hours without power is a far more extreme situation, due to heat buildup. The average response time to restore electrical power is approximately 2.5 hours, which indicates moderately high resilience. However, during a June 29 to July 7, 2012 derecho event, over 100,000 customers in DC had power interrupted for a combined total of more than 3.6 million hours, an average of 34.28 hours per customer. This high value is indicative of low resilience to climate change.

Energy planning in the District indicates high resilience. PJM Interconnection (the regional transmission operator for the District and surrounding area) uses a rigorous planning process that assesses the impacts of sea level rise on power generation facilities. Municipal managers in DC also draw on data from past experiences with extreme weather events to assess the effects of these events on oil and gas availability and pricing.

Energy services are at risk if extreme weather events negatively affect other District services, particularly transportation. In the event of a severe storm, PEPCO relies heavily on DDOT and emergency response personnel to reopen roads so that they can repair any damage to the electrical system.

As shown in Figure 8, the majority of the qualitative and quantitative indicator data plot in the "monitor for changes" quadrant (high resilience/high importance). Ensuring a constant supply of electricity is a critical need, and the District has developed emergency planning and procedures to restore power as quickly as possible, accordingly.

D.2.2.3. *LAND USE/LAND COVER*

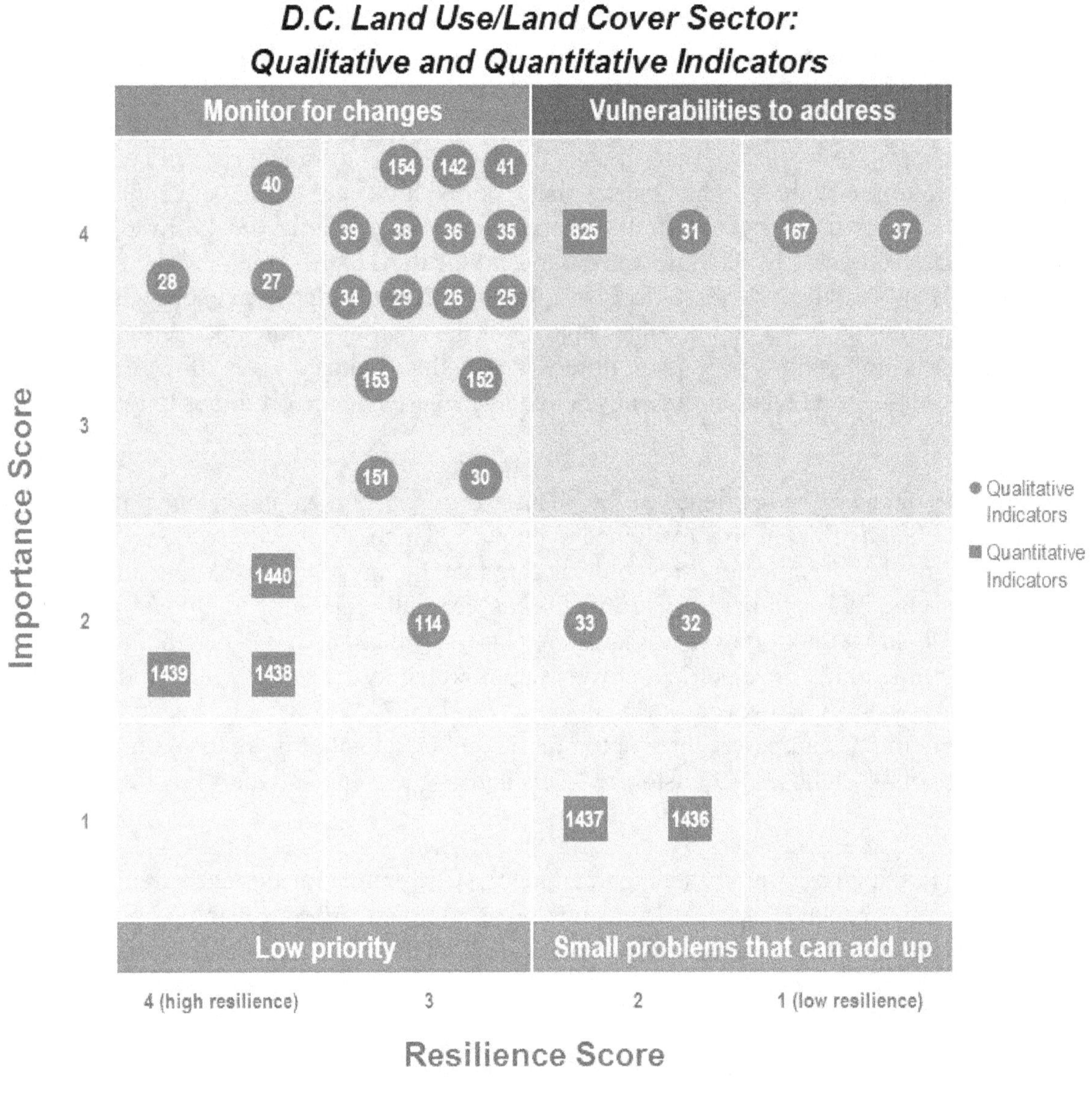

Figure 9. Washington, DC land use/land cover sector: Qualitative and quantitative indicator quadrant mapping.

Washington, DC demonstrates relatively high resilience related to land use or land cover. While important and high-value infrastructure and natural areas are located in areas vulnerable to flooding, the District has been proactive with land use/land cover planning, maximizing the benefits of urban forms; reducing heat island effects and impervious surfaces; and implementing green infrastructure and retrofits. As with most sectors, however, recognition of the importance of resilience planning and adaptation in the context of land use and land cover, and the degree of proactive response, vary across the District's neighborhoods.

The District is influenced by tides, and areas of the District along the Potomac and Anacostia Rivers and the Tidal Basin are at sea level. Therefore, the District is vulnerable to flooding and impacts of sea level rise. The District also saw a 0.19% increase in impervious cover between 2001 and 2006.

While the percentage of the District's population living in the 500- and 100-year floodplains (2.5 and 1.6%, respectively) is relatively low, the monetary value of infrastructure in the 500-year floodplain is high, and natural areas are highly vulnerable to flooding.

Only a small percentage of open/green space is required for new development, although the requirement varies across the District. While residents place high importance on green space and the District is requiring more public spaces to be green and/or pervious, increasing green space is difficult in high-density areas. Developers are also reluctant to accommodate more green space, as nearby National Park Services land is easily accessible to residents, and more than 90% of the District is within a 10-minute walk of green space.

However, the District received general high resilience ratings in areas related to proactive planning and sustainable development. The National Capital Planning Commission (NCPC) works with the DC government on federal areas in the District and has a shared comprehensive plan that includes sustainability policies.

The District is developing efforts to use urban forms to mitigate climate change impacts and maximize the benefits of urban forms. However, the degree of implementation varies across the District, and there is little focus on where in the District these initiatives are taking place.

Tree cover is considered very important from an economic perspective and for livability, and there are mechanisms to support tree-shading programs in the District. Tree-planting efforts have been fairly robust and successful, although the same cannot be said for tree preservation efforts. Again, there is disparity in these efforts across neighborhoods.

The District and the National Park Service have inventoried land use/land cover types and these data will be used in planning. There are also requirements in place for retrofits in development on vulnerable land. Workshop participants noted that resilience in DC is mostly structural, rather than from wetlands and buffers. For example, many federal buildings in the floodplain have structural protections against flooding. Furthermore, there are codes to prevent development in flood-prone areas, although existing requirements are not always followed. Executive Order 11988 requires federal agencies to avoid building in floodplains to the extent possible, but Congress ultimately decides where buildings are placed in DC. For example, the site of the National Museum of African American History and Culture is in the bottom of the watershed and will need extensive protection against flooding. Several new requirements have also been proposed but not passed, including restrictions on high-hazard users (such as dry cleaners) or vulnerable populations (such as daycares) in floodplain areas.

In cases where flooding occurs, the District encourages and provides resources for rebuilding with more flood-resistant structures and methods, although regulations regarding rebuilding communities impacted by floods have not been enforced.

There are numerous existing incentives and requirements designed to reduce the amount of impervious surface, prevent development in floodplains, and increase the use of green infrastructure for stormwater management. Incentives and requirements for the last item include a green roof rebate (for new development with green roofs and adding green roofs to existing structures), the RiverSmart Homes program, stormwater requirements, impervious surface removal rebate (on water/sewer bills), impervious surface fees, and the Green Area Ratio (which considers green walls and other items in addition to green roofs). The Green Area Ratio and stormwater requirements take many factors into consideration, including habitat corridors and use of native and/or low-water-use plant species.

Green infrastructure maintenance is covered to some extent by private parties (for example, rebate recipients are required to maintain their installations). However, not all green infrastructure programs require follow-up to ensure the infrastructure (and its benefits) are being maintained.

The District also uses current and historical data, local academic research, stakeholders, and other resources (including coastal hazard maps with 1-meter altitude contours) for planning purposes and to better understand the impact of climate change on the area.

Figure 9 includes the majority of question and indicator data in the "monitor for changes" quadrant (high resilience/high importance), indicating that the land use/land cover sector overall has high resilience to climate change in relation to the qualitative indicators workshop participants found to be important.

D.2.2.4. *NATURAL ENVIRONMENT*

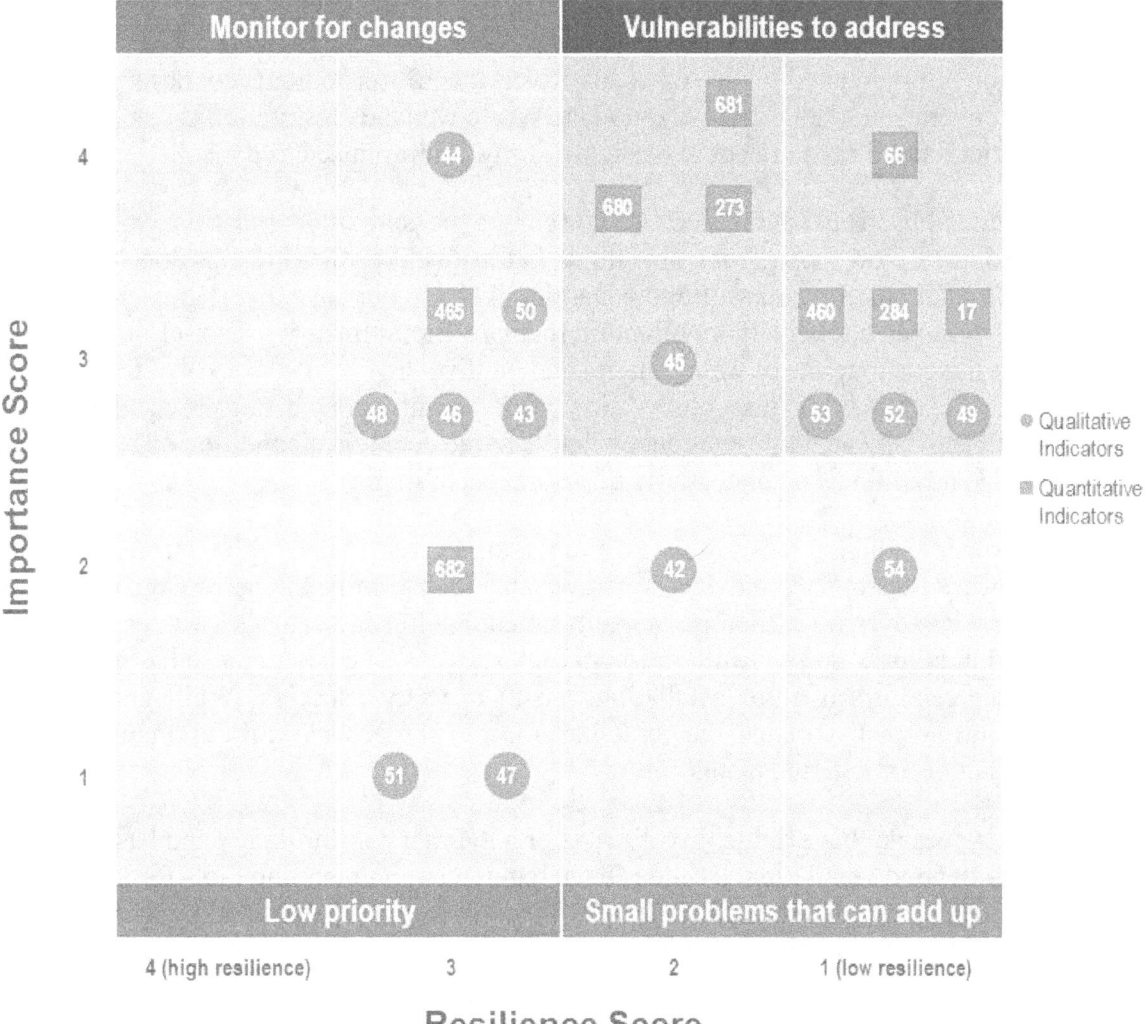

Figure 10. Washington, DC natural environment sector: Qualitative and quantitative indicator quadrant mapping.

The District's natural environment sector demonstrates low resilience with respect to the condition and status of freshwater ecosystems, physical habitat, and undeveloped land. The District has also conducted minimal planning with respect to open and green districts and ventilation. Planning in other areas related to the natural environment has been relatively robust however, and plant species diversity and the use of native plants in green infrastructure installations indicate resilience. Workshop participants assigned limited importance to the availability of environmental/ecosystem resources in situations where other District services are affected by climatic events or changes.

While no data were provided to quantify the extent to which freshwater ecosystems have been altered, workshop participants noted that less than 10 percent of the area's original wetlands likely still exist, and the Anacostia and Potomac Rivers have been channelized. While there are some conservation efforts related to the Anacostia River, the efforts do not account for substantial land coverage in the District, and no large-scale wetlands restoration projects are in place. The District's flood capacity could be significantly compromised as a result.

While the calculated Physical Habitat Index (PHI) score, a measure of degradation, was relatively high (62.31), workshop participants think that it was likely too high, as none of the sites considered are urban. Streams within the District likely do not have a PHI greater than 20. The District also received a relatively low Benthic Index of Biotic Integrity score (1.56), indicating low resilience in terms of water quality and biodiversity. Close to 19% of total species in the District are of "greatest conservation need." Although no data were available, workshop participants indicated low resilience relative to the ecological condition of undeveloped land and ecological connectivity of natural ecosystems.

Plant species diversity is high relative to the size of the urban area. Most of the introduced flora are naturalized but not disruptive. No data were available to determine the percentage change in disruptive species; however, workshop participants noted that this indicator is a vulnerability to address, as noted in Figure 10 (low resilience/high importance). The District also has native species lists, and green infrastructure installations mostly use native species. While green roofs cannot use only local or native plants, the guidelines for rain gardens and infiltration practices are to use local, native, or regional plants.

While the District does not have air quality districts or a thermal comfort index and has not analyzed areas with good ventilation, DC does have regulatory and planning tools for air quality, water quality, and land use. In addition, air quality is more strongly determined by local sources (not distant sources) and is therefore easier to control.

The District has conducted air quality analyses, implemented water protection plans, and is currently working on invasive plant protection plans. The District also plans to increase open and green space, although there may be no additional capacity for natural space in the urban area. No plans are in place to reclaim a developed area and turn it into green or open space.

Much like the land use sector, this background helps explain the relatively widespread distribution of scores in Figure 10, although the majority of data points lie in the "vulnerabilities to address" sector. This underscores the District's relative low resilience with respect to qualitative and quantitative indicators that the workshop participants found to be important.

D.2.2.5. *PEOPLE*

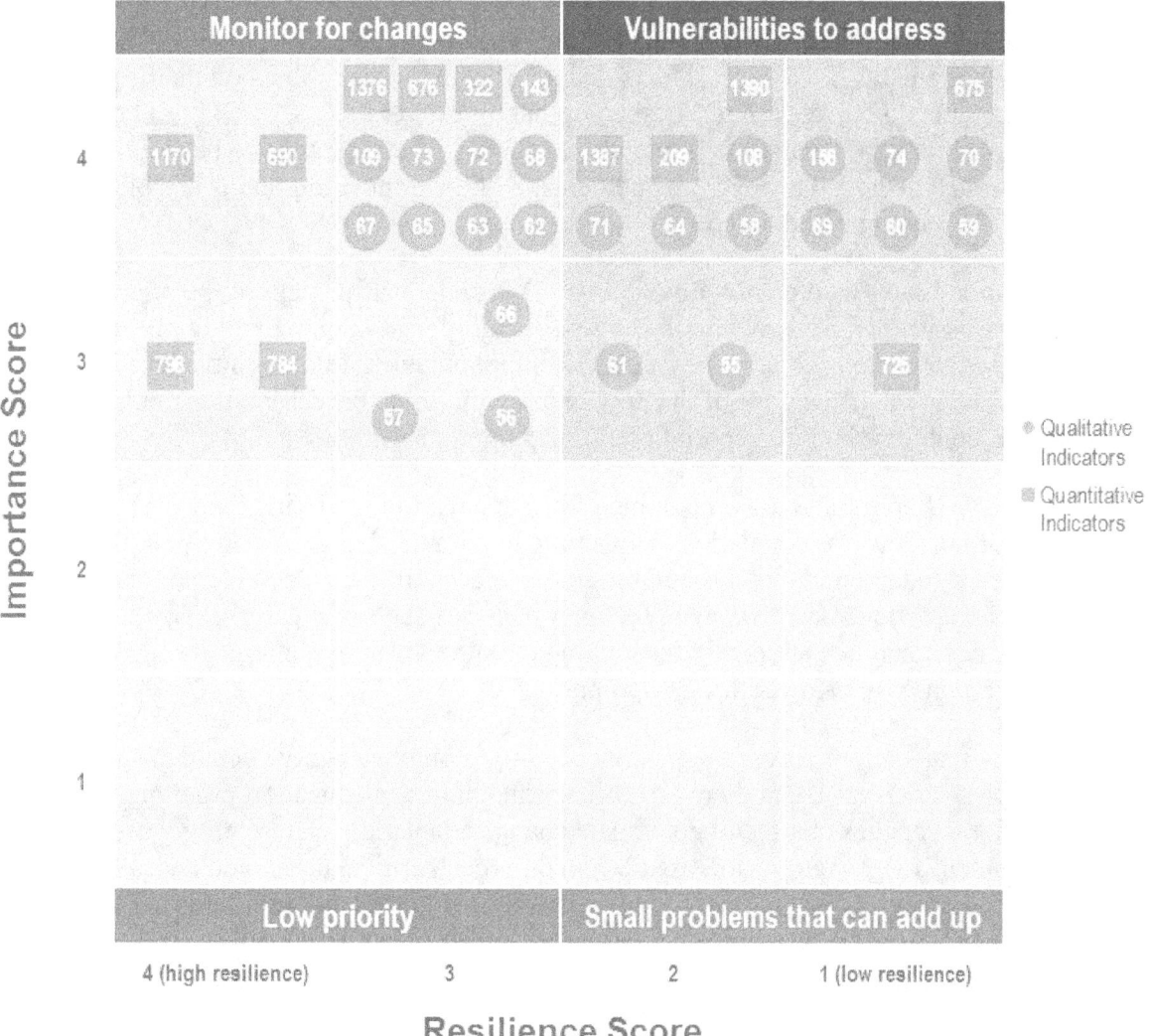

Figure 11. Washington, DC people sector: Qualitative and quantitative indicator quadrant mapping.

The District is moderately resilient in terms of its population. Similar to other sectors, there is intracity variation in resilience, and vulnerable subpopulations might be more negatively affected by climate change impacts. It is unclear to what extent ongoing outreach has impacted these populations.

Interconnectivity issues are a particular concern for this sector. The success of medical and fire responses depends on a functioning telecommunications sector, and the availability of fuel and food supplies is critical in a state of emergency. Water sector vulnerabilities can also have a significant and potentially devastating impact on public health, while health care services are heavily reliant on the energy sector. Transportation is critical for evacuations during a state of emergency.

The District is less resilient in terms of the segments of the population that are particularly vulnerable to the impacts of climate change, including the population affected by asthma (15.2% of the adult population and 22.7% of residents under 18, with these numbers on the rise), the population vulnerable due to age (16.9% of the population is over the age of 65 or under the age of 5), and the portion of the population living alone (4.3%). The latter tend to be elderly and economically disadvantaged. However, the percentage of the population that is disabled is relatively low (11.4%) compared to the 2014 national average of 22.5% (CDC, 2016).

Population location is also an area of vulnerability. Two and a half percent of the population lives within the 500-year floodplain, which is a small, high-density area. In addition, only some modes of transportation are accessible to vulnerable subpopulations (although the District scored fairly well in terms of the percentage of the population with limited access to transportation due to vulnerabilities and for whom transportation failures might be life-threatening).

To date, there have been limited to no planning efforts related to identifying demographic characteristics or locations for populations vulnerable to climate change. In addition, the District has not evaluated its adaptation policies and programs to account for vulnerable populations, although workshop participants recognized the importance of such evaluations. While some emergency services are aimed at quickly responding to vulnerable populations during power outages, these responses are slower than is optimal.

However, some organizations across the District actively promote adaptive behaviors at the neighborhood or District level, and there are policies and outreach/education programs to promote behavioral changes that facilitate climate change adaptation. These programs, driven by both the government and private sector, are designed to reach critical urban audiences. At the same time, workshop participants questioned whether these policies and programs are designed and implemented in ways that promote the health and well-being of vulnerable populations.

The District is also resilient in terms of the number of emergency responders (the number of police officers in the District is equivalent to 0.60% of the 2011 3-year American Community Survey population) and average emergency response time for fire and emergency medical service services (EMS). Over 98% of fire response times are less than 6.5 minutes, and the average EMS response time (average between fire response times and medical emergency response times) is 4.7 minutes. However, the robustness of emergency response capabilities is dependent on the resilience of the telecommunications sector. In addition, the District received a low resilience rating for the number of M.D. and D.O. physicians per capita (0.0018 active patient care primary physicians per capita).

The capacity of existing public health and emergency response systems is already limited and would not be sufficient under more extreme conditions. Likewise, the current distribution of public health workers and emergency response resources is not appropriate for the population that would be affected during an extreme event. Planning and training for response to extreme events have also been limited, both for emergency response staff and the general population (the most vulnerable populations in particular). The District might not have sufficient capacity to provide public transportation for emergency evacuations, and to date, planning for this possibility has been limited. Early warning systems are in place (including television and phone alerts) for meteorological extreme events, but these systems rely on the individual to heed the warnings and instructions.

Current rates of waterborne disease, heat-related deaths, and infectious disease are relatively low (0.02% impacted, 0.0002% of deaths, and 1.34% of the population impacted, respectively). Workshop participants noted that heat-related deaths are likely underreported and infectious disease rates are skewed by sexually transmitted disease rates. In terms of avoiding or responding to heat-related illness, the District is resilient. The District has multiple evacuation and shelter-in-place options available to residents in the event of a heat wave, and already has robust programs in place for providing public access to cooling centers, although broader efforts to reduce heat island effects could still be implemented.

The District is likewise resilient for infectious disease response. Public health agencies have identified infectious diseases and/or disease vectors that might become more prevalent in the urban area under the expected climatic changes and have developed associated response plans to reduce the associated morbidity/mortality. However, the healthcare community is not necessarily prepared for the changes in treatment necessitated by climate change and has insufficient funding to do so. For example, the District currently does not have the appropriate staff for West Nile virus surveillance.

Figure 11 shows a wide and relatively even distribution of responses across the resilience axis, indicating that while the District has made strides to address the effect of climate change on the District's population, work still needs to be done. In addition, workshop participants identified all qualitative and quantitative indicators relating to the effect of climate change on the population of the District as of high importance; no qualitative nor quantitative indicator was ranked below the median for importance.

D.2.2.6. TELECOMMUNICATIONS

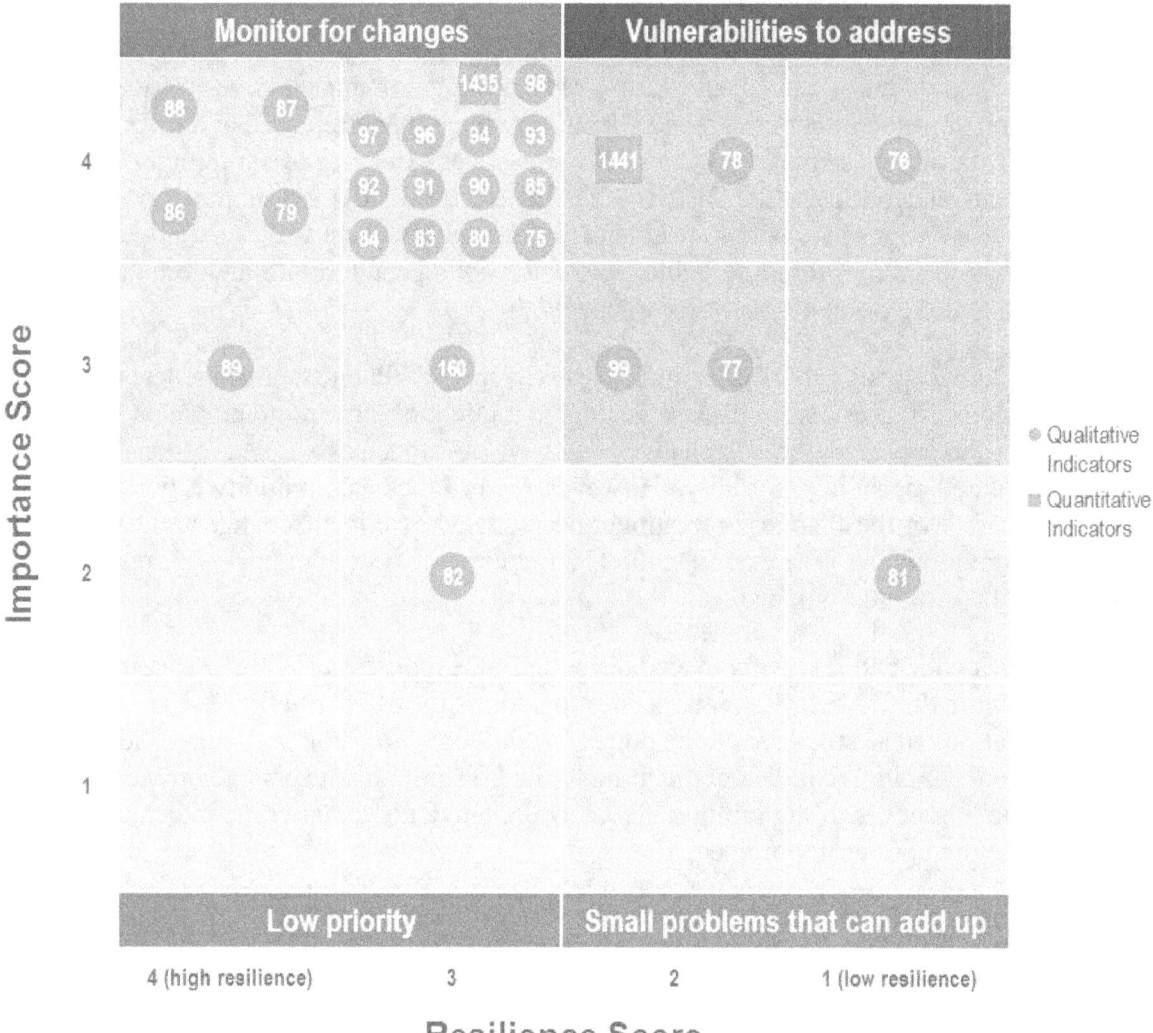

Figure 12. Washington, DC telecommunications sector: Qualitative and quantitative indicator quadrant mapping.

Washington, DC demonstrates fairly high resilience in the telecommunications sector. While loss of telecommunications infrastructure could have a significant economic impact, the District's emergency systems, emergency preparedness, and ability to maintain a communications network during an extreme event are strong. The District demonstrated more limited resilience in its ability to transmit key messages and information to residents (indicating increased vulnerability for the people sector).

The greatest areas of vulnerability identified by workshop participants include the likelihood of temporary loss of telecommunications infrastructure having a significant impact on local economies, regional economies, and the population's access to FEMA emergency radio broadcasts. In addition, the District's 9-1-1 service has no backup centers outside of the District, only across different sections of the District. The District also has key nodes in the telecommunications system, the failure of which would severely affect the District's service.

However, the District's telecommunications infrastructure appears relatively resilient to the gradual impacts of climate change or extreme climatic events. Few belowground infrastructure components are vulnerable to expected rises in groundwater levels or salt water intrusion, and few aboveground infrastructure components are vulnerable to expected winds. There is also a backup tower network in the event that satellite-based communications are disrupted by wet weather.

One data center was shut down and moved due to flooding concerns. During previous extreme weather events and other natural disasters, the District's services were either unaffected or only mildly affected. There is a great deal of redundancy built into the emergency communications systems and the infrastructure has capacity for increased public demand in an emergency, although staffing for 9-1-1 services is limited (there are more phone lines than staff members to answer them). The District also has access to backup 9-1-1 networks that could handle the majority of the load for the main emergency response networks if necessary. However, telecommunications systems do not have sufficient water and energy supply to handle more than a small amount of the anticipated extra load in the case of sudden natural disasters.

The District does not have concerns regarding the vulnerability of the telecommunications infrastructure to high temperatures or prolonged high temperatures, as long as there is power to provide the necessary cooling (demonstrating interconnectivity between the energy and telecommunications sectors).

Communications and links across infrastructure service providers and between local authorities and the service providers are good, and stakeholders can quickly make and implement decisions in emergency situations. District planners have also consulted with other city governments with similar telecommunications systems to learn how those governments coped with natural disasters and to plan for similar events accordingly.

There is some concern that the availability of telecommunications resources could be impacted if other District services, particularly power, were impacted by climatic events or changes. Backup power for these resources is provided, although the extent to which backup power is provided by diesel generators is unclear. The District has 72 hours' worth of diesel, so emergencies that extend beyond that time frame pose a greater risk to this and other sectors.

This background supports the distribution pattern for telecommunications in Figure 12. Seventy-four percent of the qualitative (20 of 27) and quantitative indicators fall in the "monitor for changes" segment (high resilience/high importance).

D.2.2.7. TRANSPORTATION

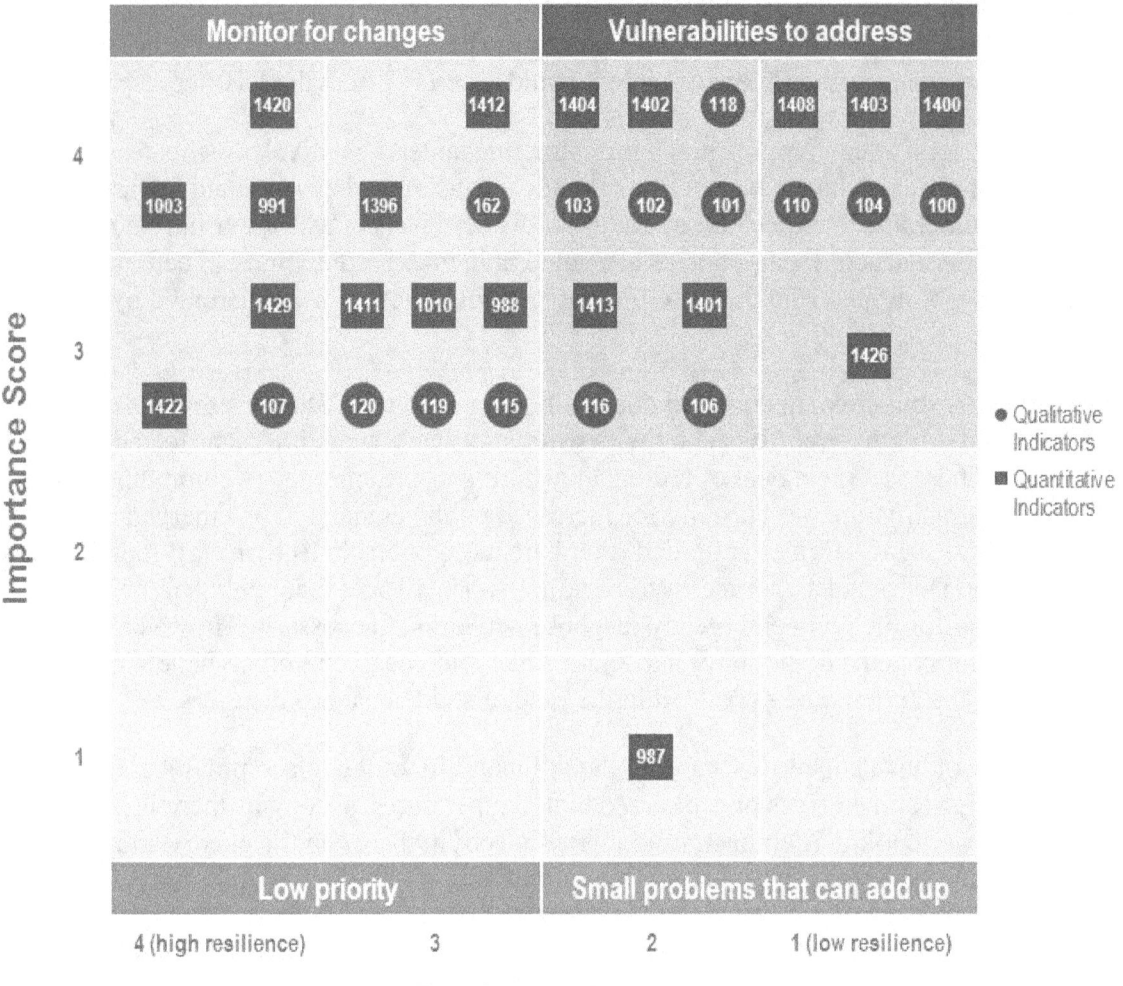

Figure 13. Washington, DC transportation sector: Qualitative and quantitative indicator quadrant mapping.

Washington DC demonstrates high resilience in terms of the level of accessibility, variety of public transportation, and travel-time scores, as well as the District's livability and walkability, although some less dense areas of the District are not as walkable. Workshop participants noted that if these indicators focused on range of livability or walkability across neighborhoods, the District would have received lower resilience scores in these areas.

The District is also generally considering climate change adaptation and resilience for transportation planning and has implemented measures to some extent. However, the current transportation infrastructure, particularly the Metro, is not equipped to handle either the gradual impacts of climate change or impacts of extreme climatic events, and limited or no funding is available to remedy this issue. Transportation infrastructure is particularly vulnerable to flooding (both in terms of the impact on transportation availability and infrastructure, as well as stormwater management) and heat, and recovery from a major climatic event could be complex and lengthy.

The District's transportation system is highly flexible, and residents have access to seven modes of transportation (via land, water, and air). Eighty-two percent of residents are near a transit stop, and the District has high levels of transport diversity and intermodal passenger connectivity. Most Metro stops provide access to bus connections, and the average distance of non-work-related trips is fairly short (under 5 miles, although this might vary across District neighborhoods). While the mean travel time to work for residents in the District is high, 29.6 minutes compared to the national average of 25.4 minutes, workshop participants assigned this indicator a low importance ranking. In addition, the District ranks highly in terms of number of telecommuters or potential telecommuters. Roadway connectivity is also high, but workshop participants noted that a high number of intersections could also increase the likelihood of accidents and the amount of road and traffic light maintenance required.

The District has taken proactive steps to develop and implement resilience-building approaches and incorporate climate impact considerations into transportation projects, alongside reactive disaster response plans. The District also has a severe weather plan. In terms of infrastructure, the District has tested new or innovative materials that might be more capable of withstanding the anticipated impacts of climate change, and the District has planned for green infrastructure and requires its implementation. However, workshop participants noted that while green infrastructure planning has occurred, the plans have not necessarily been executed. District agencies have also been working to upgrade bridges and update evacuation and road/bridge infrastructure planning to consider extreme climate events.

The District also received a high resilience score for the annual congestion costs saved by operational treatment costs, calculated at $53 per capita. The District has high-occupancy vehicle lanes and procedures for clearing traffic accidents from bridges. During traffic incidents, DC and Virginia can quickly change grid patterns to keep traffic moving, although whether these actions truly alleviate congestion has not been determined.

Despite planning efforts, the District's transportation infrastructure is still highly vulnerable and not equipped to handle the gradual impacts of climate change or the devastation that a severe event could bring. Workshop participants assigned low resilience scores for resistance of major transportation links and critical nonroad transportation facilities to the anticipated impacts of climate change.

It is unclear whether flooding would significantly affect critical facilities. Ten percent of critical roadway and rail line miles are within the 500-year floodplain, and depending on the data used, either 5.6 or 11% are within the 100-year floodplain. Workshop participants noted that rain can hit the District quickly and heavily, causing vents and tunnels to flood. District culverts are not sized to meet future (or even current) stormwater requirements, but upgrades will be completed by 2030. In addition, 31 bridges (12.8% of District bridges) are structurally deficient.

In addition to flooding, increased temperature places considerable stress on the District's transportation infrastructure. Few materials currently used in the District's transportation systems are compatible with anticipated temperature changes, but many of the District's transportation systems were built for the climate as it was at the time. When metro rails overheat, they develop heat kinks, requiring the District to replace that part of the rail. If a kink goes unnoticed, trains will derail. However, communication procedures are in place to prevent risks associated with heat kinks.

Congestion is also an issue for the District. One study ranked the District first in the nation for yearly delay per auto commuter[5] among the very large urban areas[6] in 2014 (Schrank et al., 2015), although workshop participants questioned the validity of the study. Another study ranked the District third for congestion intensity and second for congestion costs (Litman, 2016).

The District is developing and implementing plans to replace aging infrastructure, but not all of these plans account for the anticipated impacts of climate change. Funding for infrastructure repair and replacement is also limited and very competitive. The District currently has no funding mechanisms specifically for adapting transportation systems to climate change.

In terms of emergency response and recovery, residents are generally unaware of evacuation procedures, and the length of time required to restore major high-traffic vehicle transportation links in the urban area after a failure could be significant but would vary depending on the scenario. Even now, a short-duration problem on the Metro causes significant travel delays, and the Metro system has very limited redundancy. One of the District's current goals is to increase modal redundancy (for non-climate-related reasons). The District is adding bike lanes and streetcars, and it hopes to improve the Metro's redundancy and increase the bus system's flexibility to reduce the impact of incidents on the Metro system.

Finally, the transportation sector is relatively reliant on other sectors, particularly energy, to remain operational. Availability of transportation resources is generally at significant risk if climatic events or changes affect other District services. Likewise, short- or long-term problems in the transportation sector would significantly impact other sectors, particularly people and economy. The District is also relatively dependent on the long-range transportation of goods and services.

Figure 13 indicates that the transportation sector in Washington, DC has significant vulnerabilities to climate change impacts and shows a wide and relatively even distribution of responses across the resilience axis. This indicates that while the District has made strides to address the effect of climate change on the District's population, risks remain high. In addition, workshop participants identified all qualitative and quantitative indicators relating to the effect of climate change on the District's transportation sector to be of high importance. Only one indicator regarding travel time to work (which, as noted, is above the national average in the District) ranked below the median for importance.

---

[5] Yearly delay is defined as extra travel time during the year divided by the number of people who commute in private vehicles in the urban area.
[6] Areas with over 3 million population.

D.2.2.8. *WATER*

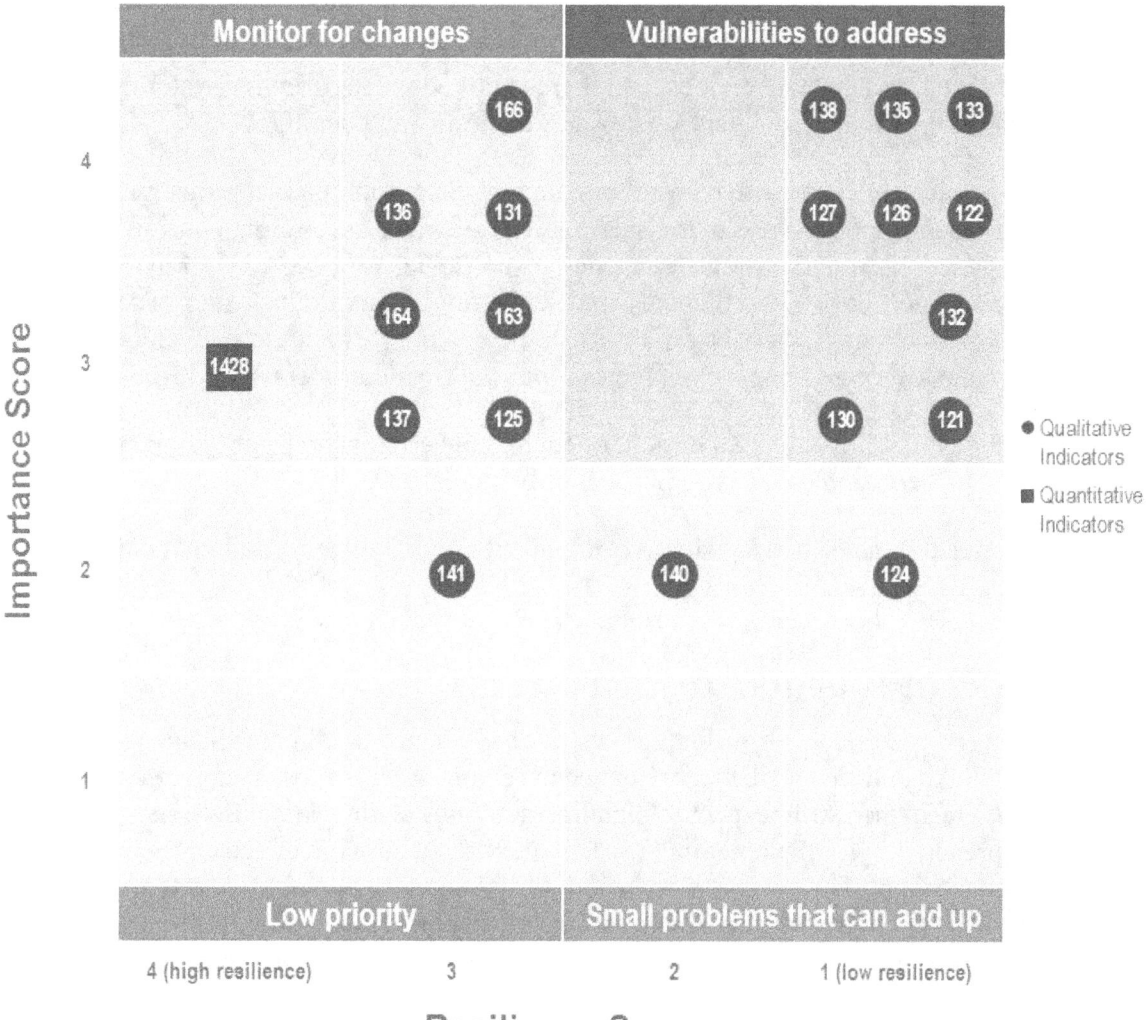

Figure 14. Washington, DC water sector: Qualitative and quantitative indicator quadrant mapping.

The District's water sector exhibits low resilience to climate change, particularly with respect to source and infrastructure (but exclusive of planning activities). Interconnectivity with other sectors is important, as disruptions to water service could significantly impact public health and the economy, land use/land cover, and the natural environment. The water sector is also heavily dependent on the resilience of the energy and transportation sectors.

The Potomac River is the only source of drinking water for the District. Additionally, there are few interconnections with neighboring water systems. While water quality is sufficient in terms of numbers of Safe Drinking Water Act violations, almost the entire Potomac watershed is outside Washington, DC, limiting the District's control over water quality. Moreover, there is no treatment to handle increases in nutrient loading.

Water infrastructure is at high risk during extreme events. More than 90% of stormwater and wastewater pump stations are in the flood zone. Minimal backup power is available for drinking water, stormwater, and wastewater services, and there is no redundant drinking water treatment system. Likewise, there are no redundant wastewater or stormwater services.

The water sector is more resilient with respect to planning. The drinking water treatment plant has redundant chemical suppliers, and there is a hierarchy of water use protocol during a shortage or emergency. A water/wastewater agency response network (WARN) provides technical resource support during emergencies, and storm sewers and drains to storm sewers have been inventoried, although there is variability in the extent to which these inventories inform planning (in part because one single agency does not own the stormwater infrastructure).

Drought and water availability are not a current or future concern for the District; the District anticipates that its present ample water supply will only increase.

Figure 14 shows that the majority of qualitative and quantitative indicators are ranked as important.

D.2.3. SUMMARY OF WASHINGTON, DC FINDINGS

The results of the DC case study show that the District's resilience to climate change is mixed, with areas of both high and low resilience within each sector. Across most sectors, the District demonstrated high resilience with respect to planning activities and a general awareness of the need to prepare proactively for the potential impacts of extreme climatic events or the gradual impacts of climate change. The District also benefits from an existing, robust transportation system; network of parks and other green spaces; a relatively small size; and uniqueness in terms of federal government presence and involvement. The role of the federal government in the District perhaps grants it more expertise and resources for climate readiness and emergency preparedness than other metropolises of a similar size might receive.

However, in some areas—especially transportation—the city's current infrastructure is less resilient, particularly to the impacts of flooding or rising temperatures, and the resources to make needed improvements are unavailable. In addition, the resilience scores across all sectors might not accurately reflect significant intracity disparities in resilience. Workshop participants frequently noted that disparities in economy, infrastructure, transportation access, and population vulnerability could mean that climate change disproportionately affects some areas of the District more than others. It is also unclear to what extent programs and messages regarding climate change and adaptation and emergency response reach the most vulnerable subpopulations.

APPENDIX E. WORCESTER, MA CASE STUDY

This appendix contains the Worcester, MA case study. Section E.1 provides background on the known climate vulnerabilities faced by Worcester and on the existing planning the city has undertaken to address these vulnerabilities. Section E.2 reviews the results for Worcester, MA. Results are by sector and accompanied by visual data summaries.

E.1. WORCESTER, MA BACKGROUND

Worcester, MA is a postindustrial city. Like many of its counterparts across New England, the East Coast, and the Midwest, Worcester faces challenges in finding the resources to sustain critical infrastructure, health services, and human services for current needs, let alone the resources to prepare and incorporate responses to the threats posed by climate change.

Worcester is the second largest city in New England. The city is located in central Massachusetts, approximately 45 miles (72 kilometers) west of Boston, the state capital and largest New England city. Its current population is near 185,000. Like many cities in the North and Midwest, its population peaked in the 1950s at just over 200,000 during the immediate postwar years. After decades of decline (in accordance with national trends), population growth became positive in the 1980s and is projected to remain so; total city residency increased by 5% between 2000 and 2010 (WRRB, 2013).

Similar to other postindustrial cities, Worcester has faced the challenge of reinventing and revitalizing its economy. While the city grew and prospered from the mid-1800s through World War II driven by thriving textile, metalworking, and machine tool industries, it faced economic decline through the second half of the twentieth century. However, the city has seen some economic recovery in recent years from growth in the biomedical/life sciences, health services, and higher education sectors (City of Worcester, 2004; WMRB, 2008), similar to other large "rust belt" cities, such as Buffalo and Cleveland (populations approximately 258,000 and 390,000, respectively; U.S. Census Bureau, 2015a; U.S. Census Bureau, 2015b). Recovery has not been constant; from 2001 to 2007, Worcester lost more than 2,200 jobs, or 2% of its total employment base (Boyle, 2011). The biotechnology cluster in particular has become an increasingly important anchor in the regional economy and a key component in the state's economic development initiatives (O'Sullivan, 2006). As a result, the city's employment structure has shifted; the leading employers are currently hospitals and associated medical service organizations (WRRB, 2015). Median household income remains below the national average (U.S. Census Bureau, 2013c).

Worcester has a continental humid climate, similar to many cities in the Midwest and Northeast. Continental humid climates are typified by large seasonal differences in temperatures with precipitation throughout the year (Kottek et al., 2006). Worcester is vulnerable to a range of climate extremes, from damaging ice, blizzards, and cold air events to heat waves. The city must therefore plan for a broad range of contingencies. The hilly topography surrounding the city can magnify disaster impacts and complicate recovery efforts (CMRPC, 2012). For example, there is a greater risk of water being funneled into valleys and rivers and a risk of landslides, which can disrupt transportation, telecommunications, and other sectors.

Like many small- to medium-sized cities across both New England and the United States, Worcester has undertaken relatively little climate change-related adaptation planning. The city has an existing Climate Action Plan that focuses largely on mitigation measures, such as reducing greenhouse gas emissions. The plan currently does not include a focus on adaptive measures.

Worcester is also part of the Central Massachusetts Regional Planning Commission (CMRPC), a group focused on planning and responding to natural disasters. However, this group does not concentrate on understanding changes in the intensity or frequency of these hazards, nor on ways to adapt to such changes. Partially due to the lack of planning, data are also limited. Worcester is therefore a good test for the tool in a data- and planning-limited environment. Additionally, Worcester is more typical of many American cities, in that it does not receive high levels of support and coordination for many city functions from the federal government, unlike the District.

E.1.1. KNOWN VULNERABILITIES

Worcester is located in a continental humid climate, with year-round precipitation and large seasonal temperature fluctuations, making the city vulnerable to climate extremes on both ends of the spectrum. The city is vulnerable to flooding and severe storms (including hurricanes, Nor'easters, and winter storms, with associated flooding and high winds), as well as extreme cold, ice-damming of rivers, extreme heat, and urban fires (CMRPC, 2012). Table 13 lists several historic weather events that have impacted the city of Worcester.

Stormwater flooding, aggravated by urban runoff, is especially prevalent; Worcester accounts for nearly half of historical claims in the region for damages related to stormwater flooding. Riverine and dam flooding are also concerns. Worcester contains six dams considered "high hazard" and four deemed "critical" by the Office of Dam Safety. The 100-year floodplain in Worcester contains several critical facilities, including a fire department and three medical clinics. On the other hand, Worcester has received higher marks than any other community in the region from the National Flood Insurance Program for its aggressive program to raise awareness of flood hazards and maintain elevation certificates on new and improved buildings (CMRPC, 2012).

Storms with high winds and winter storms can cause power outages that may threaten vulnerable populations. Extreme cold and extreme heat are also public health and safety concerns, especially with regard to the city's homeless population. Although the *Central Massachusetts Region-wide Pre-Disaster Mitigation Plan* (CMRPC, 2012) does not explicitly address climate change, it is likely that Worcester's susceptibility to urban fires (there were 815 fires between 2004 and 2009) could be exacerbated by extreme heat or drought conditions.

Intense precipitation events, which are expect to increase in frequency, can place a strain on sewer and wastewater infrastructure. The city has both separate and combined sewage and stormwater systems. The oldest part of the system, a combined sewer system that covers 4 square miles, includes pipes constructed of brick in the mid to late 1800s (City of Worcester, 2013a). Changing precipitation patterns and higher temperatures could affect water quality as well. Worcester's water supplies meet all federal and state drinking water standards but are considered highly susceptible to contamination due to uncontrolled uses (i.e., activities on privately owned lands) in the 40-square-mile watershed (City of Worcester Water Operations, 2014).

Table 13. Major weather and other events and their impacts in Worcester, MA

Weather event	Date	Impacts
The Great New England Hurricane	September 1938	Structural damage and flooding were heavy. According to the Worcester *Gazette*, "Buildings were partially collapsed… roofs ripped off, church steeples toppled, store fronts [sic] blown out…chimneys leveled, signs torn down and the streets littered with glass..." (Herwitz, 2012). Severe flash flooding also affected the city and surrounding areas, with 10–17 inches of rainfall reported for both the hurricane and storms in the preceding days (Foskett, 2013). Tree damage was so severe that a temporary sawmill set up in Hawden Park processed lumber for over 2 months (Foskett, 2013), including nearly 4,000 street trees downed (Herwitz and Nash, 2001).
Worcester Tornado	June 9, 1953	Ninety-four people were killed, more than 1,000 people were injured, and more than 10,000 people were left homeless in Worcester and the surrounding areas from an F4 tornado. The storm remains the deadliest New England tornado on record (Fortier, 2013; Herwick, 2014).
Dutch Elm Disease	1950s	Due to vase-shaped spreading crowns and the ability to grow in compacted soils, elms were widely used as street and landscaping trees. An introduced fungal blight killed virtually all of Worcester's elms during the 1950s and 1960s, significantly reducing the city's tree canopy and its capacity to take up stormwater, filter air, and provide shade during hot summers (Herwitz and Nash, 2001).
Hurricane Gloria	September 1985	Heavy winds and rain damaged trees and power lines. President Reagan declared much of New England, including Worcester, a federal disaster area (FEMA, 1985).
April Fool's Day Blizzard	April 1997	Thirty-three inches of heavy snow fell, setting the record for the snowiest April and causing extensive damage to trees and power infrastructure across Massachusetts (Rosen, 2015). The late arrival of the blizzard meant that many plows and snow-removal equipment were already in storage, making restoration of transportation networks difficult (Marcus, 1997).
Hurricane Irene	August 2011	Heavy winds and rains resulted in localized flooding and power cuts (WBUR, 2011).

Weather event	Date	Impacts
The Endless Winter	2014–2015	Over the 2014–2015 winter, Worcester recorded 119.7 inches of snow, nearly double the average of 64.1 inches. Worcester was the second snowiest city with a population over 100,000 in America, less than 1 inch below the first place city of Lowell, MA (Golden Snow Globe, 2015). The late January "Blizzard of 2015" dumped more than 34 inches on the city in 24 hours, breaking the 110-year record for the snowiest day in the city's history (Eliasen, 2015). The winter taxed transportation infrastructure; made commutes dangerous, as snowbanks obscured sightlines for pedestrians and drivers; and resulted in structural damage from rooftop ice dams.
Asian Long-Horned Beetle; Emerald Ash Borer	2000s–Ongoing	These introduced insects kill maple, ash, beech, and other common New England trees, damaging the urban forest and reducing its capacity to provide water retention, air filtration, shade, and scenic values. At present, over 25,000 trees in the Worcester area have been destroyed in order to prevent further spread of the pests (Freeman, 2009), and quarantine measures have been put in place to prevent movement of infected firewood (City of Worcester, 2015).

Residents with few resources (e.g., the poor and the homeless) may be particularly vulnerable to extreme weather events and temperature extremes. From 2007 to 2011, approximately 19% of the city population was living below the poverty level, compared to 10.7% for the state overall (U.S. Census Bureau, 2013). In January of 2013, approximately 1,202 Worcester residents were homeless, with 22 unsheltered (living on the street) (CMHA, 2013).

More than 4% of the city's roads and railroads are within the 500-year floodplain (although the city received a high resilience rating based on the percentage of roads and railroads within the 100-year floodplain). The current transportation designs and related infrastructure planning regimes are not considering impacts of climate change or resilience. The system is designed to state standards and under major budget constraints at present; if these standards were changed to consider climate change scenarios, such considerations of climate risk would be incorporated. The city recognizes the need for substantial local (and national) investment to simply repair crumbling roads and bridges, let alone increase resilience.

E.1.2. EXISTING ADAPTATION AND MITIGATION PLANNING

Worcester, like many smaller U.S. cities, is subject to the effects of climate change but has fewer resources than major metropolitan centers, such as Washington, DC, for addressing, planning, adapting, and responding to those effects. The project team reviewed information on 10 U.S. cities similar to Worcester in size; although a majority had sustainability or hazard mitigation plans addressing one or more specific areas (e.g., water supply, flood hazards), only one in 10 had been comprehensively evaluated for resilience to climate change. Cities with more developed climate change resilience evaluations and adaptation plans tended to be those that took advantage of external resources, such as the Resilient Communities for America program (RC4A, 2013). This national campaign encourages communities to formally pledge to develop more resilient cities and provides critical resources for community leaders to assist in this process. A number of community resilience organizations support the program, including: International Council for Local Environmental Initiatives (ICLEI)—Local Governments for Sustainability, the National League of Cities, the U.S. Green Building Council, and the World Wildlife Fund.

Smaller cities generally have fewer resources for evaluating resilience, studying climate threats, and planning for them. They also lack the economies of scale that larger cities can benefit from. Their smaller staff may be unable to participate as fully as those of larger cities in activities (e.g., conferences) that provide access to new information and opportunities for regional and national partnerships. Further, smaller cities may be at greater risk of losing institutional knowledge as a result of staff turnover. On the other hand, smaller cities with fewer stakeholders may find it easier to develop consensus around policies and implement them. In addition, smaller cities are generally less vulnerable to the urban heat island effect (ICLEI, 2010).

Worcester's climate planning is typical of a smaller city. To date, climate adaptation has not received sustained attention as a matter of city policy. Across all sectors, dedicated planning and initiatives geared towards increasing resilience, through emergency preparedness, improved redundancy, incorporating climate change considerations into planning, and infrastructure replacement and design, is somewhat scattered and limited. In many cases, these limitations are the result of lack of funding, although limited awareness and interest among residents and coordination across city government are also cited as issues in some sectors. The city has conducted some tabletop exercises for climate change adaptation planning, looking at past events to assess the effectiveness of current measures. Funding for specifically adaptation-related activities or measures is limited to nonexistent.

The city's Climate Action Plan (City of Worcester, 2006), focused on climate change mitigation (i.e., reducing greenhouse gas emissions), also discusses several measures that have implications for climate adaptation. For example, diversifying Worcester's energy portfolio to include more local and alternative energy sources will improve the city's ability to cope with extreme events. Specific projects in the plan include a 100-kW hydropower turbine at the water filtration plant, a 250-kW wind turbine at new North High School, and a biodiesel (B-20) pilot program at Hope Cemetery. Similarly, improved energy efficiency will reduce vulnerability to climate-related disruptions and free up resources to cope with environmental challenges. Energy efficiency measures discussed in the plan include investing in fuel-efficient vehicles, implementing anti-idling technologies and policies for city vehicles, and promoting the adoption of energy-efficient appliances. Other actions discussed in the plan, including developing a municipal green building policy (e.g., promotion of cool roofs), planting community gardens, and protecting open and green spaces, will help mitigate the urban heat island effect. The plan also calls for improved collection of energy and climate data, which can support adaptation efforts.

Detailed reporting of the city's progress toward the goals laid out in the 2006 Action Plan is not available. However, Worcester received a "Green Community" designation in 2010; it qualified for this designation by developing and implementing specific plans and policies (e.g., policies that make it easier to site and permit renewable/alternative energy projects; and plans and policies to reduce energy consumption by the municipal government or the city as a whole).

Worcester participates in the CMRPC, which has developed a predisaster mitigation plan that addresses multiple hazards (CMRPC, 2012). Although the plan does not specifically discuss climate change, it addresses a number of hazards where climate change might exacerbate the frequency or intensity, including flooding, severe winter storms, Nor'easters, hurricanes, tropical storms, drought, extreme heat, and extreme cold. The plan discusses actions, potential funding sources, priority listings, and proposed schedules to address these hazards. Proposed actions include:

- Identifying and prioritizing structural mitigation projects, such as stormwater drainage and dam repair
- Adding catch basins at low elevation points and upsizing some existing basins
- Performing a hydraulic analysis at strategic locations
- Constructing a relief surface sewer
- Evaluating and repairing dams in the city
- Cleaning/managing stormwater structures and basins
- Increasing communication/coordination between all government, municipal, private, and nonprofit agencies regarding predisaster mitigation
- Helping residents build working relationships with the utility company to improve communications during events.
- Implementing/improving hazard warning systems and notifications to vulnerable populations.
- Developing educational and outreach tools to reach marginalized populations.
- Integrating disaster mitigation concerns into projects for various sectors, including transportation and land use.
- Planning for capital needs.
- Collaborating with other interested parties to identify predisaster mitigation activities.

The Commonwealth of Massachusetts's Executive Office of Environmental Affairs (2004) has developed a Lower Worcester Plateau Ecoregion Assessment that covers an area including some of Worcester and several neighboring communities. Mindful of the benefit of forests in moderating climate (among other benefits), the assessment's authors recommend several actions (e.g., providing incentives, producing educational materials, and assessing valuation studies) to help protect forestland in the region from development.

The city's ability to respond effectively to an extreme climatic event rests upon its existing services, and discussions suggest that the city's emergency response capabilities are relatively strong. Based on planning measures in place today, it is unclear whether the city will be appropriately prepared in the long term to manage the impacts of gradual climate change across all sectors. However, the findings suggest that the city, with appropriate resources and focus, can incorporate adaptation and resilience considerations into existing plans, practices, and programs.

E.1.3. DATA COLLECTION APPROACH

The primary data collection approach for the Worcester case study was discussions with key individuals who have knowledge and experience in each of the eight sectors (see Appendix B). Two Clark University faculty members oversaw a team to identify the most appropriate individuals within the city of Worcester to provide feedback on the urban resilience tool's qualitative and quantitative indicators. In addition, a literature search was conducted on the city of Worcester. Relevant literature was reviewed for background information, as well as information on key metrics for Worcester.

First, all participants discussed the relevance of each qualitative indicator. Then, they provided an importance weight for each qualitative indicator. Finally, participants were asked to identify the best score for each qualitative indicator from the question and options provided. The process was repeated for the quantitative indicators, also requesting that participants discuss relevant available data sets to determine the value of each indicator and review a threshold-based resilience score (if provided).

The project team spoke with at least one primary individual with in-depth knowledge for each of the eight sectors (see Appendix B). In the case of data analysis for the water and people sectors, the project team recorded the response of the individual whom they deemed most qualified and knowledgeable regarding the specific qualitative or quantitative indicator.

E.2. Worcester, MA Results

Figures 15 and 16 highlight overall trends in the Worcester, MA data. For both resilience and importance, scores ranged from 1 to 4, with 1 indicating lowest resilience or lowest importance, and 4 indicating highest resilience or highest importance. In Figures 15 and 16, the "resilience score" represents an average score for all qualitative or quantitative indicators in that sector. These sectors are ranked, from left to right, by the average importance score for that sector. As such, a sector with a low resilience score towards the right of the plot may be considered relatively vulnerable compared to another sector with a low resilience score towards the left.

Note that there were no responses to questions for the people sector (i.e., no qualitative indicators), and no data in the energy and telecommunications sectors for quantitative indicators.

In general, Figure 15 shows minimal spread in the qualitative indicator data for resilience. Average importance scores for all sectors are also clustered. With the exception of the energy sector's average resilience measurement, on average no other sector scores below the median for resilience or importance. However, importance scores are almost always higher than resilience scores, suggesting potential across-the-board vulnerability. These data suggest that the most sectors have similar levels of vulnerability, with the transportation, land use/land cover, and energy sectors the least resilient on average.

For the quantitative indicator data in Figure 16, there is a wide range of scores. This is similar to the wider range of quantitative indicator scores compared to qualitative indicator scores reported for the District (see Figures 3 and 4). These data suggest that the natural environment and water sectors are the most resilient sectors for which there are data. The other sectors (transportation, people, economy, and land use/land cover) had similar importance scores and much lower resilience scores, suggesting that these sectors may need more attention. All average resilience scores for quantitative indicators (see Figure 16) are higher than the resilience scores for qualitative indicators (see Figure 15).

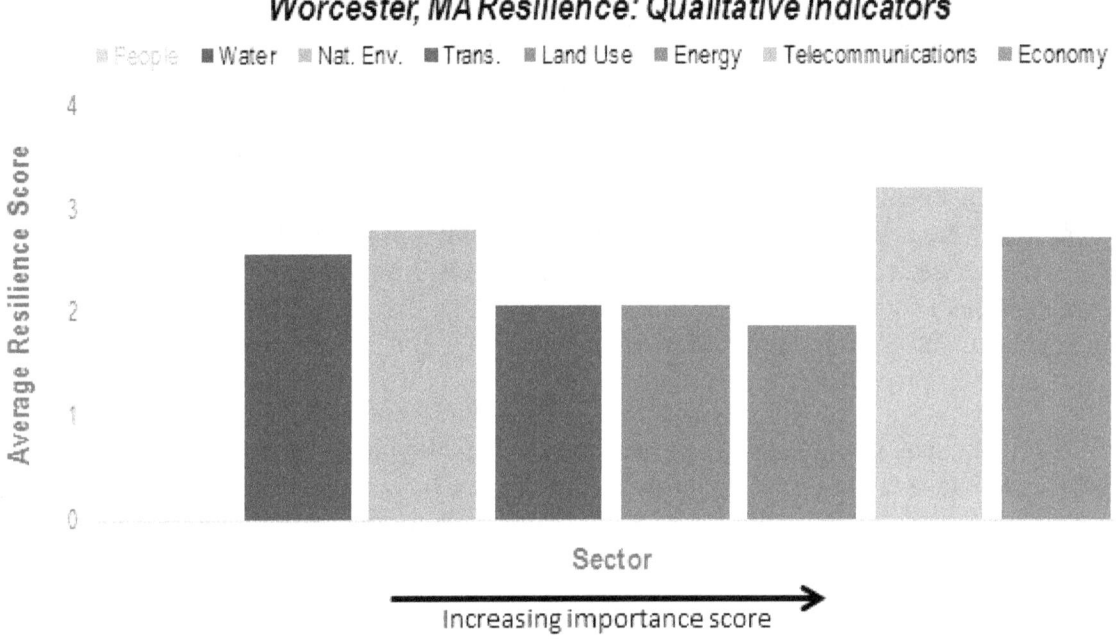

Figure 15. Worcester, MA: Average qualitative indicator resilience and importance.

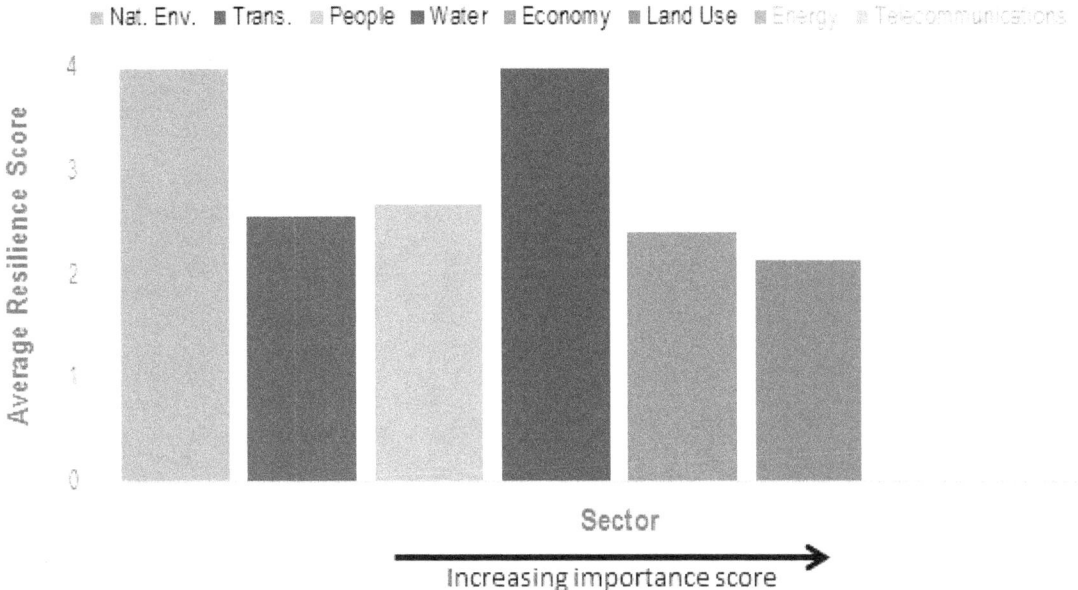

Figure 16. Worcester, MA: Average quantitative indicator resilience and importance.

Figure 17 disaggregates the data summarized in Figure 15 and highlights potential "spikes" of high risk within sectors with overall lower averages. For example, qualitative or quantitative indicators associated with the water sector fall in all four main quadrants of Figure 17, and all seven sectors with data in Worcester have at least one qualitative indicator appearing in the "vulnerabilities to address" domain.

Data collected in response to questions as qualitative indicators for Worcester cluster in the "monitor for changes" quadrant (slightly more than 60% of the total). Of the seven sectors with data, only economy is overwhelmingly restricted to the "monitor for changes" quadrant (9 out of 11 qualitative indicators). This suggests that most sectors in Worcester need to pursue a variety of strategies to adequately prepare for climate change, as well as prioritize actions.

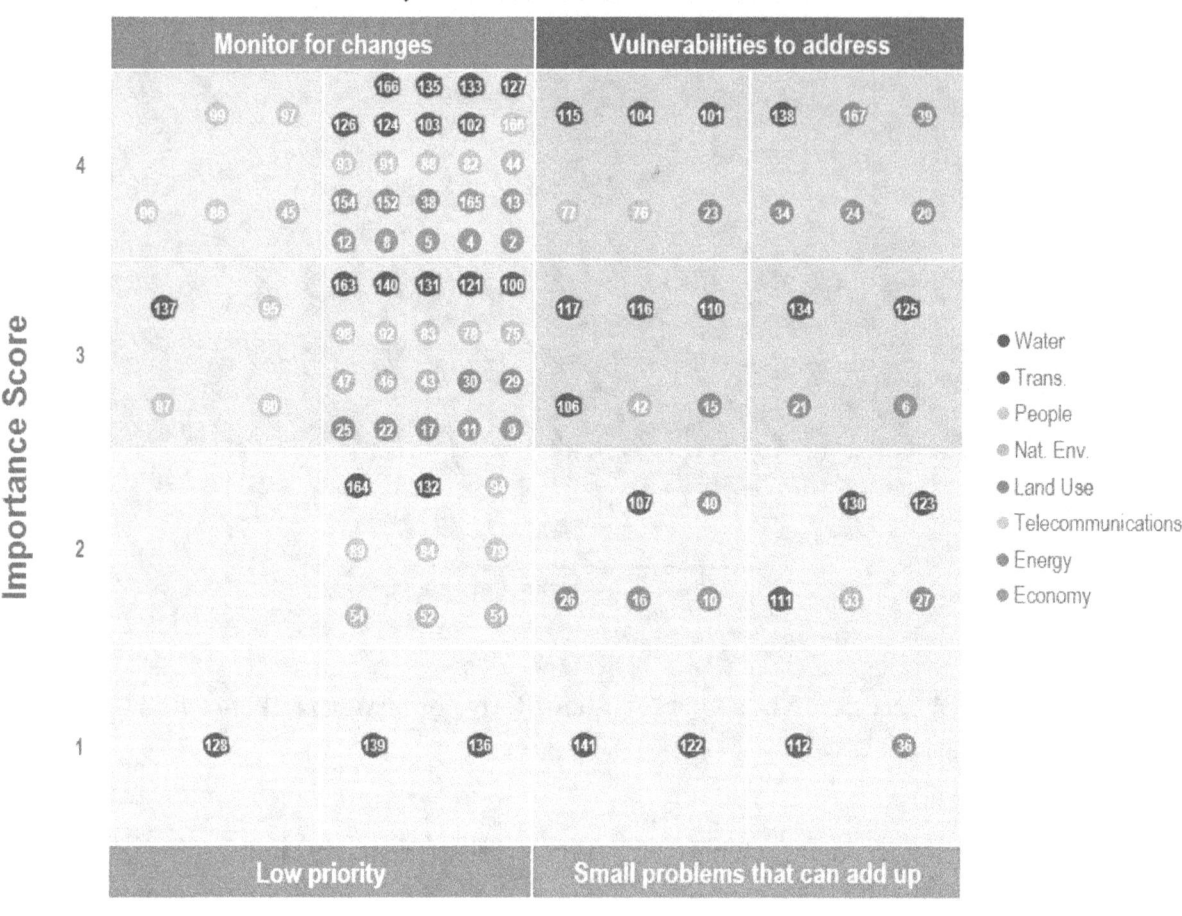

Figure 17. Worcester, MA: Qualitative indicator quadrant mapping.

Figure 18 offers the same presentation as Figure 17, but for quantitative indicator data across all sectors. Note that no quantitative indicator data were available for the energy or telecommunications sectors. Much like the qualitative indicator data in Figure 17, the majority (four out of six) of the sectors with available data have at least one indicator in the "vulnerabilities to address" quadrant, highlighting how averages can hide specific facets of climate preparedness that cities need to address.

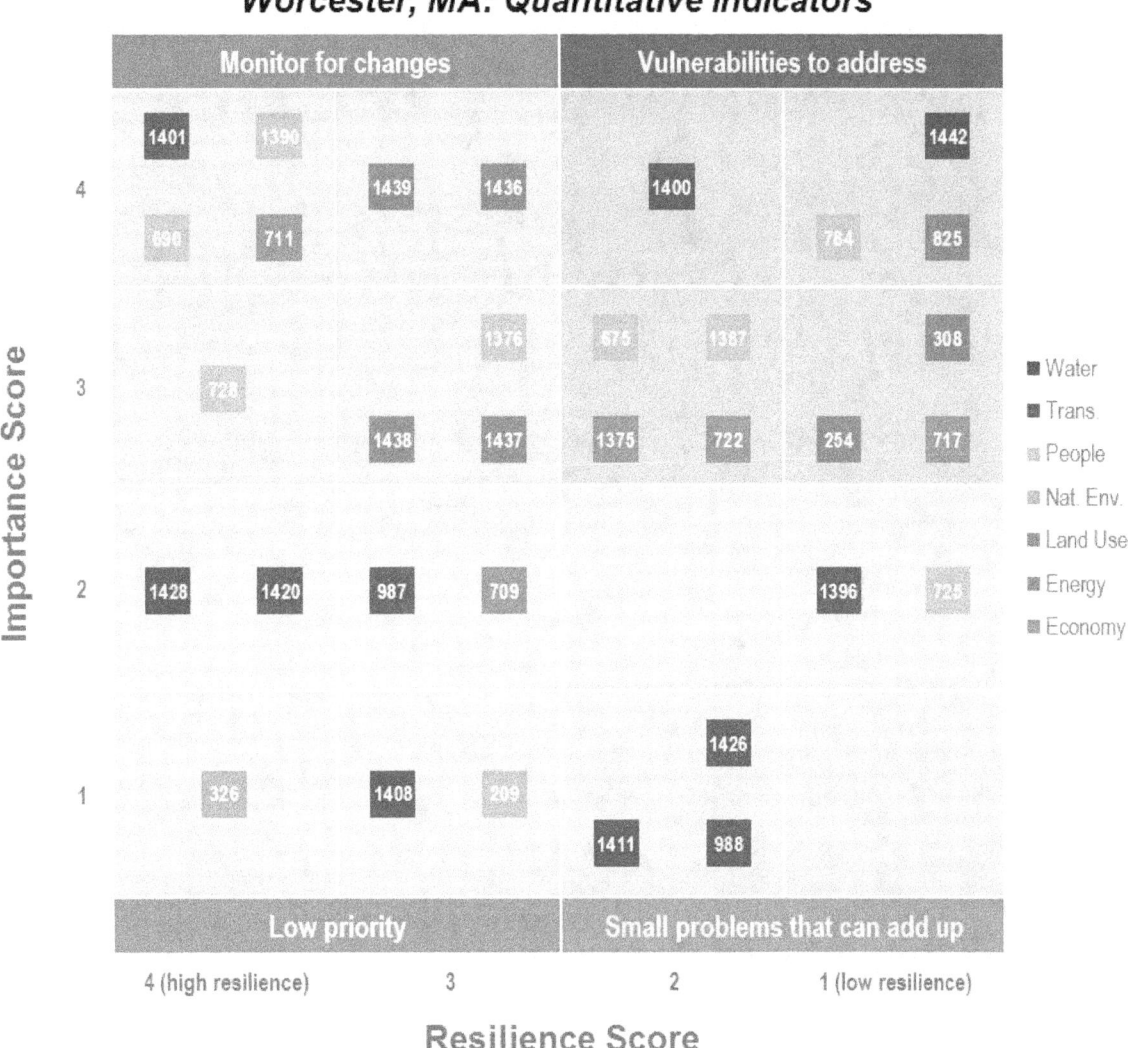

Figure 18. Worcester, MA: Quantitative indicator quadrant mapping.

E.2.1. SECTOR-SPECIFIC INVESTIGATIONS

The sections below connect the results above to potential underlying drivers and roadblocks for each sector, discussed in the literature as well as from participant input. However, unlike Washington, DC, little supplemental literature was available for Worcester. In addition, the discussions were primarily limited to one representative (as the tool was designed), in contrast to the participation of numerous representatives across sectors, as was the case at the DC workshops.

E.2.1.1. *ECONOMY*

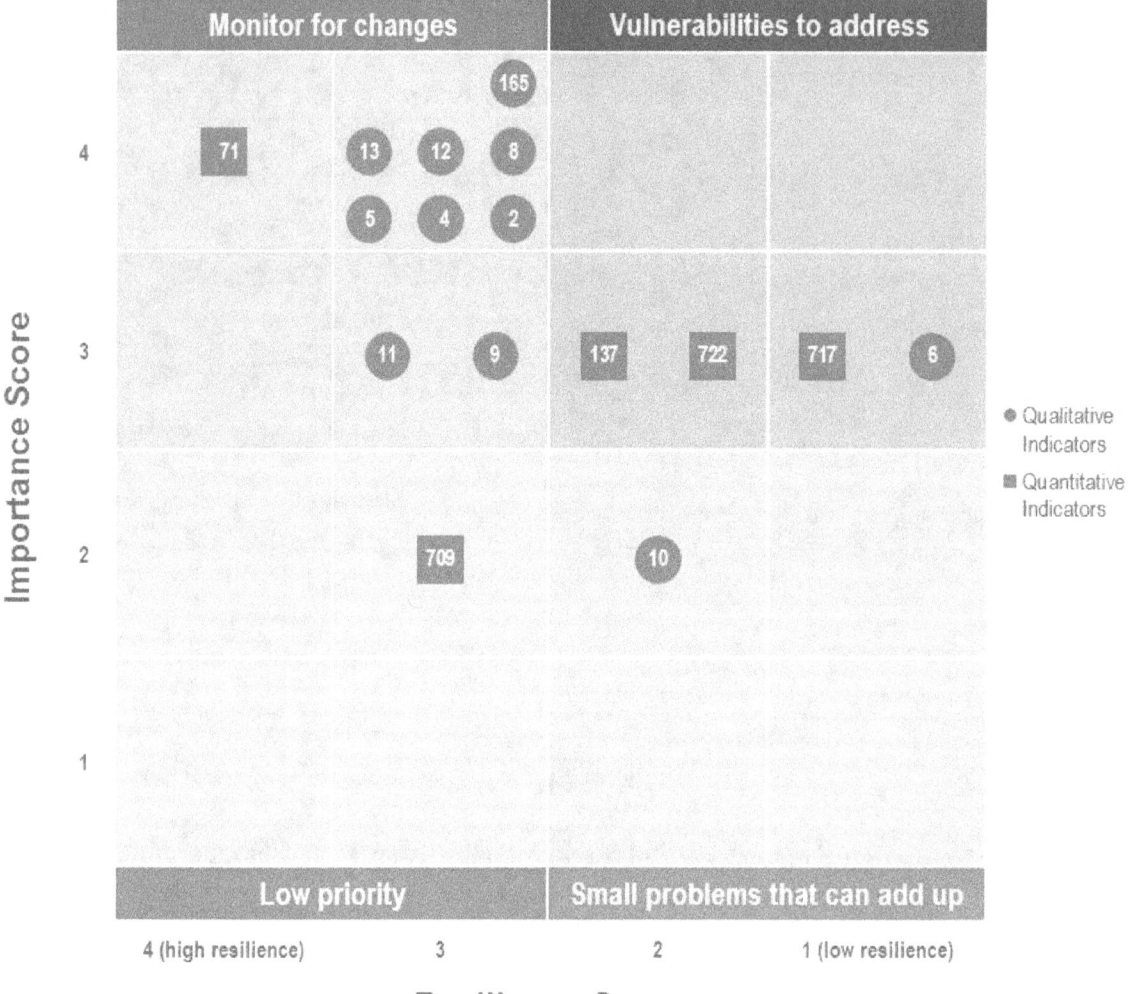

Figure 19. Worcester, MA economy sector: Qualitative and quantitative indicator quadrant mapping.

Overall, the diversity of economic options and the generally improving economic outlook for the city potentially resulted in the moderate to high resilience scores. While manufacturing still plays an important role, the education (Worcester holds over 13 colleges and universities) and health care sectors now account for nearly half of total city employment and make up the largest share of the city's economy. The city's biotechnology industry is also growing, leveraging the educated workforce and health care sectors. The city is not heavily dependent on climate-sensitive sectors such as agriculture.

Worcester received lower resilience scores for the economy due to vulnerable subpopulations, including the growing homeless population (which increased 5.1% between January 2012 and January 2013) and the percentage of the city's population living below the poverty line (19%, based on 2007–2011 data). However, the 2006 passage of the Massachusetts health care reform law required most state residents to obtain some level of health insurance coverage, and as a result, 95% of the noninstitutionalized population has access to health insurance.

In the event of climate shock, it is unclear whether jobs lost in one sector could be replaced by expanding the economy and job opportunities in another sector. The participant noted that resilience in this regard depends on which sector is disrupted and on workers' skills and mobility in each sector.

As Worcester continues to prepare and improve its economic resilience to climate change, the city may consider exploring additional funding opportunities and modifying its management approaches to climate change adaptive planning. The participant noted that adaption planning responsibilities are spread out over multiple offices within Worcester's government, reducing projects' efficiency and efficacy, and no funding is currently available for multipurpose adaptive development projects (meeting both recreation and adaptive development needs), which may be roadblocks to planning efforts.

Figure 19 shows that 69% of the qualitative and quantitative indicators lie in the "monitor for changes" quadrant (high resilience/high importance), indicating that Worcester's economy has high resilience to climate change. Worcester's "vulnerabilities to address" (low resilience/high importance) in its economy sector relate to the city's vulnerable subpopulations and gaps in adaptive planning funding. Figure 19 also demonstrates that the city may reconsider its approach to adaption planning, possibly concentrating planning activities into one office, as this issue lies in the "small problems that can add up" quadrant.

E.2.1.2. *ENERGY*

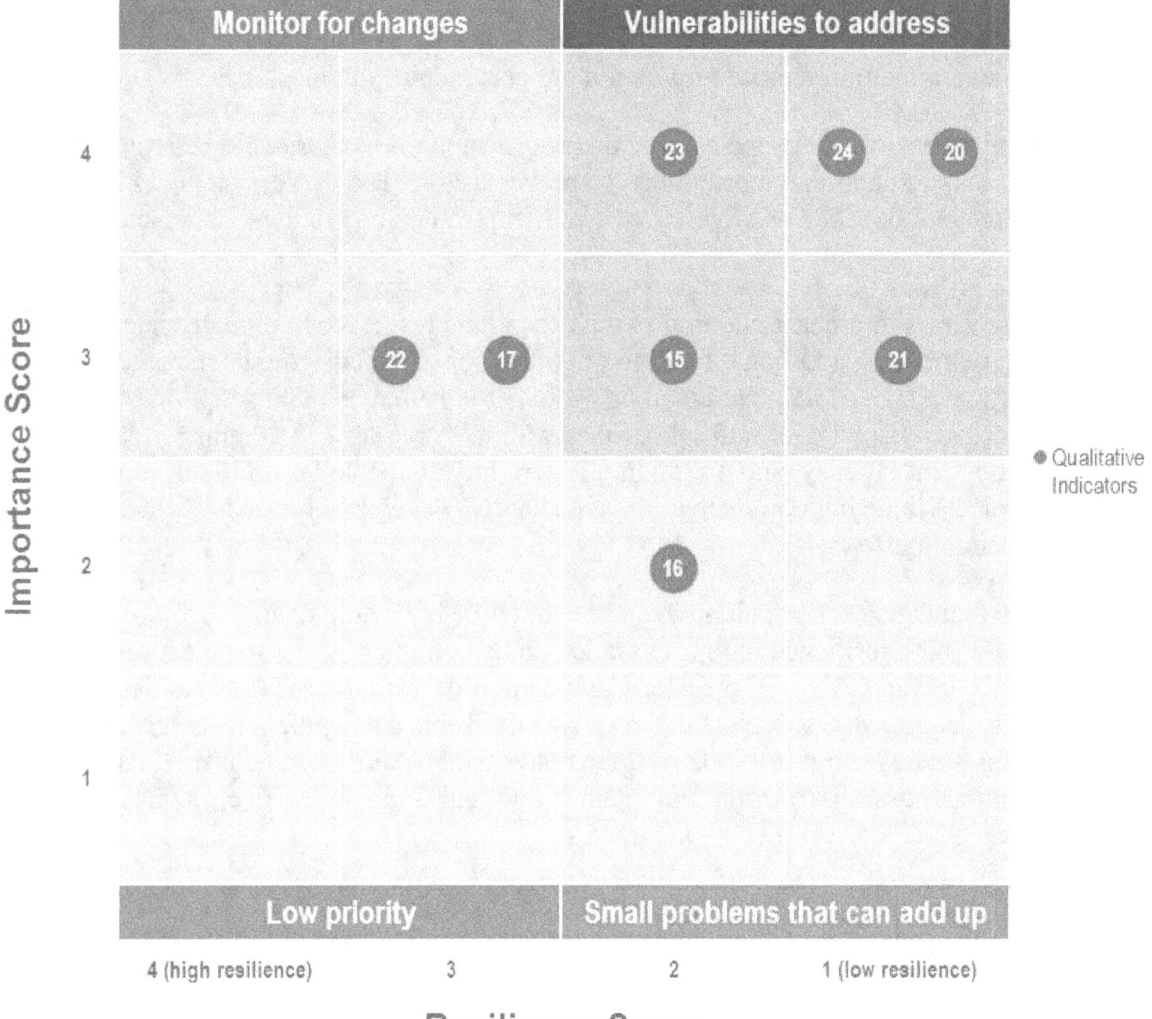

Figure 20. Worcester, MA energy sector: Qualitative and quantitative indicator quadrant mapping.

The data on the energy sector came exclusively via qualitative indicators. No indicator data were available for relevant quantitative indicators.

In terms of energy supply, the city's energy sector is relatively resilient, as energy supplies come from outside the metropolitan area to only a moderate extent. The city has also made moderate efforts to reduce energy demand. However, based on participant responses, the resilience of the city's energy sector in terms of coping with or responding to stressors (extreme events, outages, or higher peak demand/demand at different times) appears to be limited. The city's redundant energy systems have only a small capacity in the event of a threat to the energy system; however, the participant rated this as a less important factor for the city. Additionally, the response time to restore electrical power after a major event may take more than a day, and the electrical generation capacity cannot handle higher peak demands or peaks at different times than it currently experiences; the participant ranked these issues with high and moderately high importance, respectively.

Diverse and local sources of power and heat contribute to a city's resilience. Worcester relies on a combination of electricity, natural gas, and oil. Several renewable energy sources contribute to the local power grid, including a wind turbine located at a city high school (McCauley and Stephens, 2012).

Beyond efforts to encourage energy consumption reductions, the city is not actively pursuing alternative approaches to better manage demand or reduce risk through distributed generation or smart grid technologies. At the same time, the city is aware of an increased frequency of extreme events that threaten its electricity systems and the potential benefits of moving towards decentralized systems, including more distributed renewable generation. However, the participant did not believe that increased decentralization would necessarily reduce vulnerabilities to climate change. One participant made note of National Grid's smart grid pilot project in Worcester; however, based on the current status of this pilot program, the opportunities it presents for managing demand in Worcester are not particularly significant. Therefore, these areas of vulnerability may remain in the longer term.

Problems were also noted in acquiring relevant energy usage data for Worcester, especially because energy consumption data are recorded by distribution circuit, which does not match community or city limits. In addition, load zones in central Massachusetts are not defined by city. Worcester offers a "Worcester Energy Program" to encourage energy savings.

Figure 20 shows that of the data available, 62.5% (or five of eight) qualitative indicators lie in the "vulnerabilities to address" quadrant, indicating that there are significant steps Worcester can take to increase its energy sector's resilience to climate change.

E.2.1.3. *LAND USE/LAND COVER*

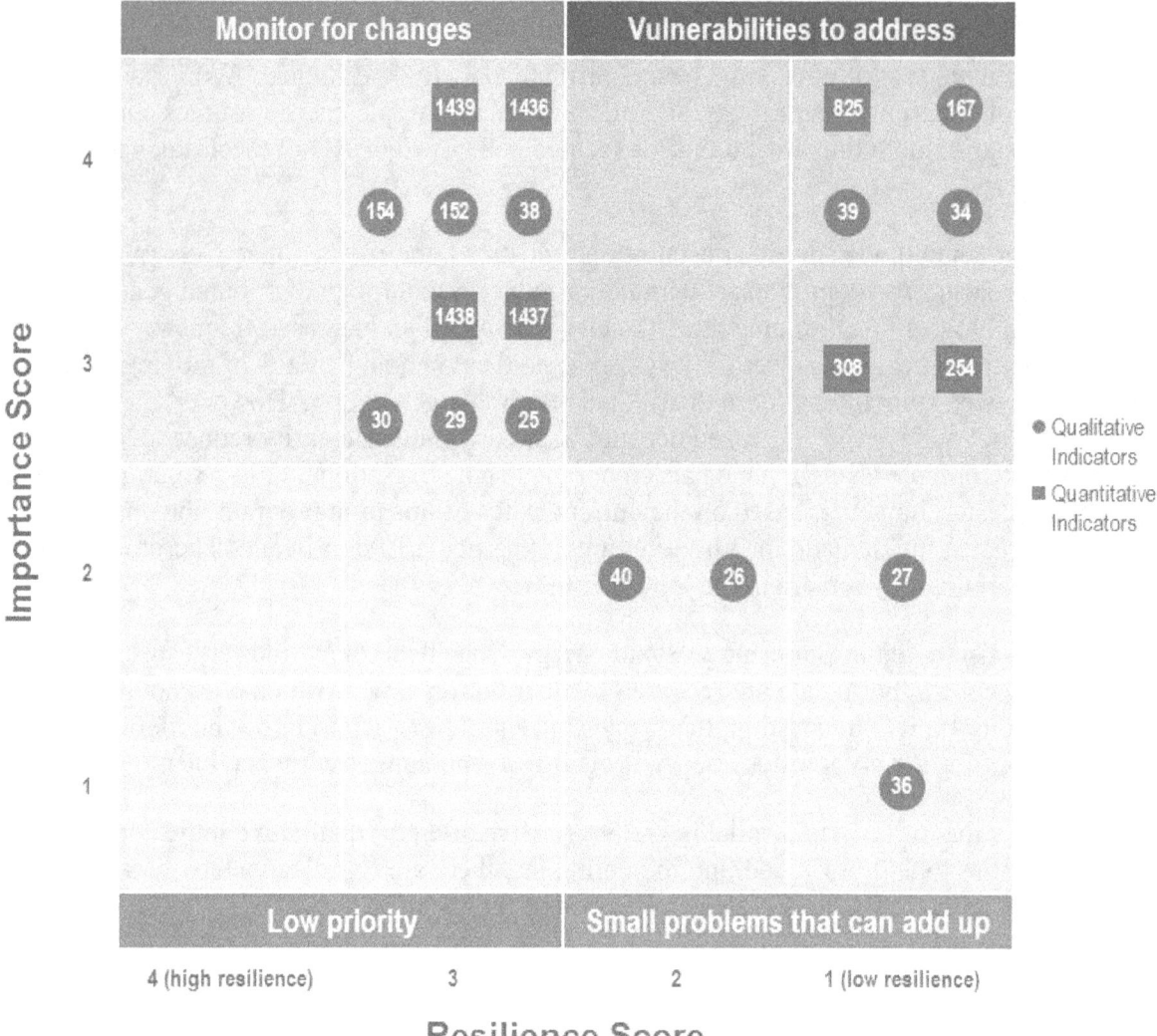

Figure 21. Worcester, MA land use/land cover sector: Qualitative and quantitative indicator quadrant mapping.

While Worcester has many planning and zoning initiatives that may be relevant to climate resilience, none of them have been justified by climate resilience or undertaken for the primary purpose of resilience. The most significant discrepancy between importance score and resilience score, indicating the greatest perceived vulnerability, concerns the location of valuable infrastructure and continued development (without concern for retrofitting) in areas that are vulnerable to extreme events, including flooding. The same discrepancy was identified regarding the lack of financial incentives to prevent development in floodplains and reduce the amount of impervious surface, among other initiatives. These responses may speak to a greater vulnerability in the economy sector than was indicated by the interview.

The city is not actively using resilient retrofits or urban forms to mitigate climate change impacts or address urban heat island effects. Although the participant placed limited importance on the latter two initiatives, the city demonstrates low resilience related to the percentage of city land that is urban (100%) and percentage of impervious cover. This underscores the need for using retrofits or urban forms to mitigate climate change impacts or address heat island effects; providing funds to reduce the amount of impervious surfaces; and limiting further development in vulnerable areas, as identified in the qualitative indicators.

Worcester is taking steps to improve the city's land use/land cover resilience to climate change. The participant noted that there are mechanisms to support tree shading programs in urban areas; however, additional funding is needed from existing, established sources within the city. Additionally, incentives exist to integrate green stormwater infrastructure into infrastructure planning to support flood mitigation. When green infrastructure was used, the participant noted that the infrastructure was selected with minimal attention to the ecological benefits provided.

However, the city has taken advantage of existing resources, including local academic research and other stakeholders, and has taken into account historical land use and land cover changes to better understand and account for climate stresses and resilience in land use planning.

Of the qualitative and quantitative indicators with high resilience, all fall into the "monitor for changes" quadrant, indicating that the participant ranked the actions the city has taken to improve land use/land cover resilience as highly important. Of the qualitative and quantitative indicators that fall in the low resilience quadrants, 60% fall into the "vulnerabilities to address" quadrant (low resilience/high importance) and 40% fall into the "small problems that can add up" quadrant (low resilience/low importance); of these qualitative and quantitative indicators, 80% have the lowest resilience score of 1.

E.2.1.4. *NATURAL ENVIRONMENT*

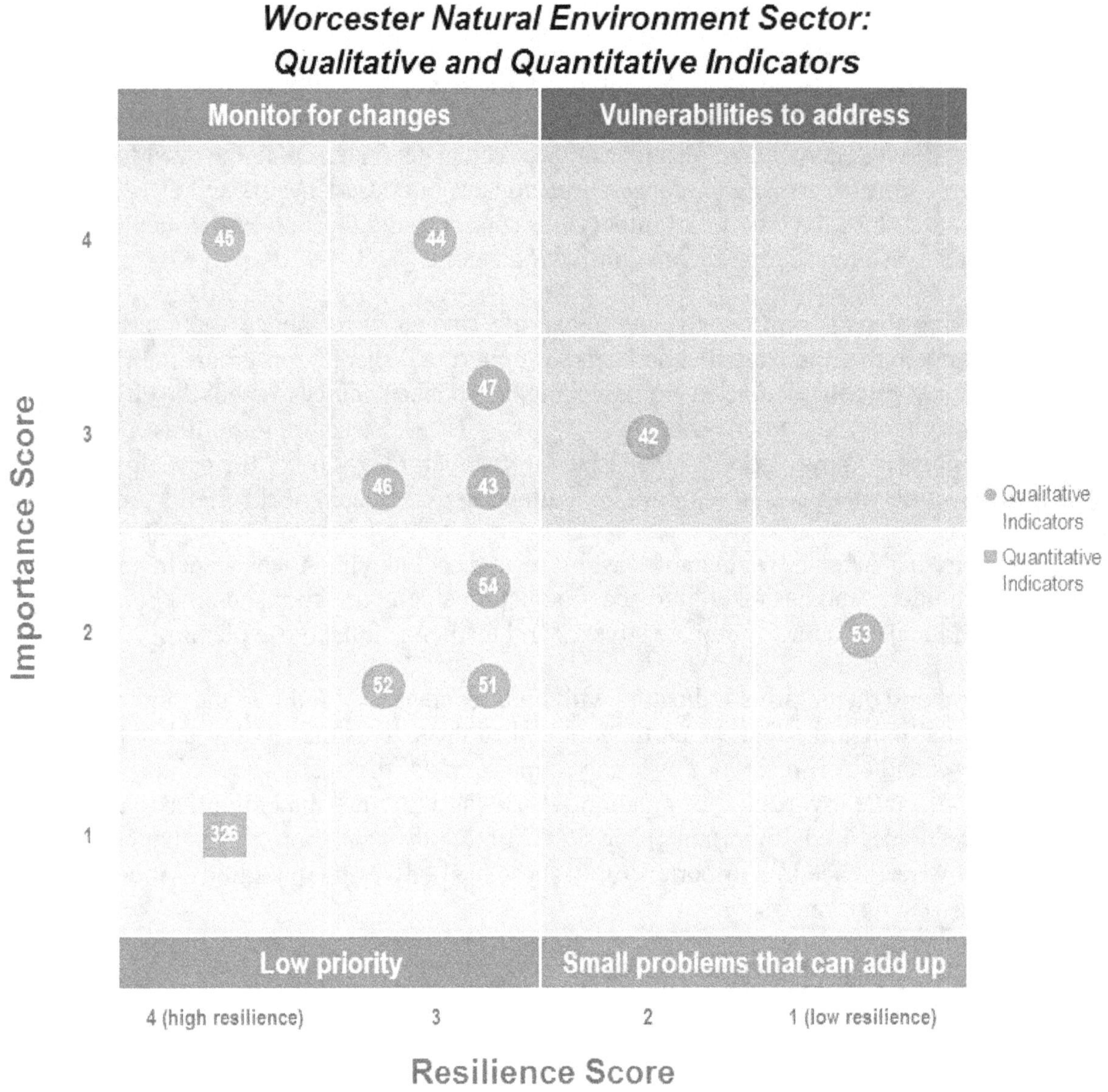

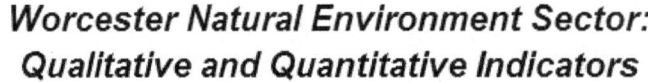

Figure 22. Worcester, MA natural environment sector: Qualitative and quantitative indicator quadrant mapping.

Worcester demonstrates relatively high resilience in this sector, based on existing regulatory and planning tools/processes on water quality, air quality, and land use; coordination with other entities on water quality issues; and green space initiatives. The city has also developed native plant and animal species lists and uses these species in green infrastructure, as seen in Figure 22, where most of the qualitative and quantitative indicators score at least a 3 for resilience.

In addition, there are few wetland species at risk (rare, endangered, or threatened). While the city has no plans for preserving areas with good ventilation, the participant assigned a lower importance score to this qualitative indicator. However, the city demonstrates limited resilience for the availability of environmental/ecosystem goods and services if other city goods and services, such as power, water, and telecommunications, were affected by extreme climate events or gradual changes. This may help explain some of the lower resilience-scoring qualitative and quantitative indicators.

Figure 22 demonstrates that Worcester has high resilience in the natural environment sector, with 82% of the qualitative and quantitative indicators having a resilience score of at least 3.

E.2.1.5. *PEOPLE*

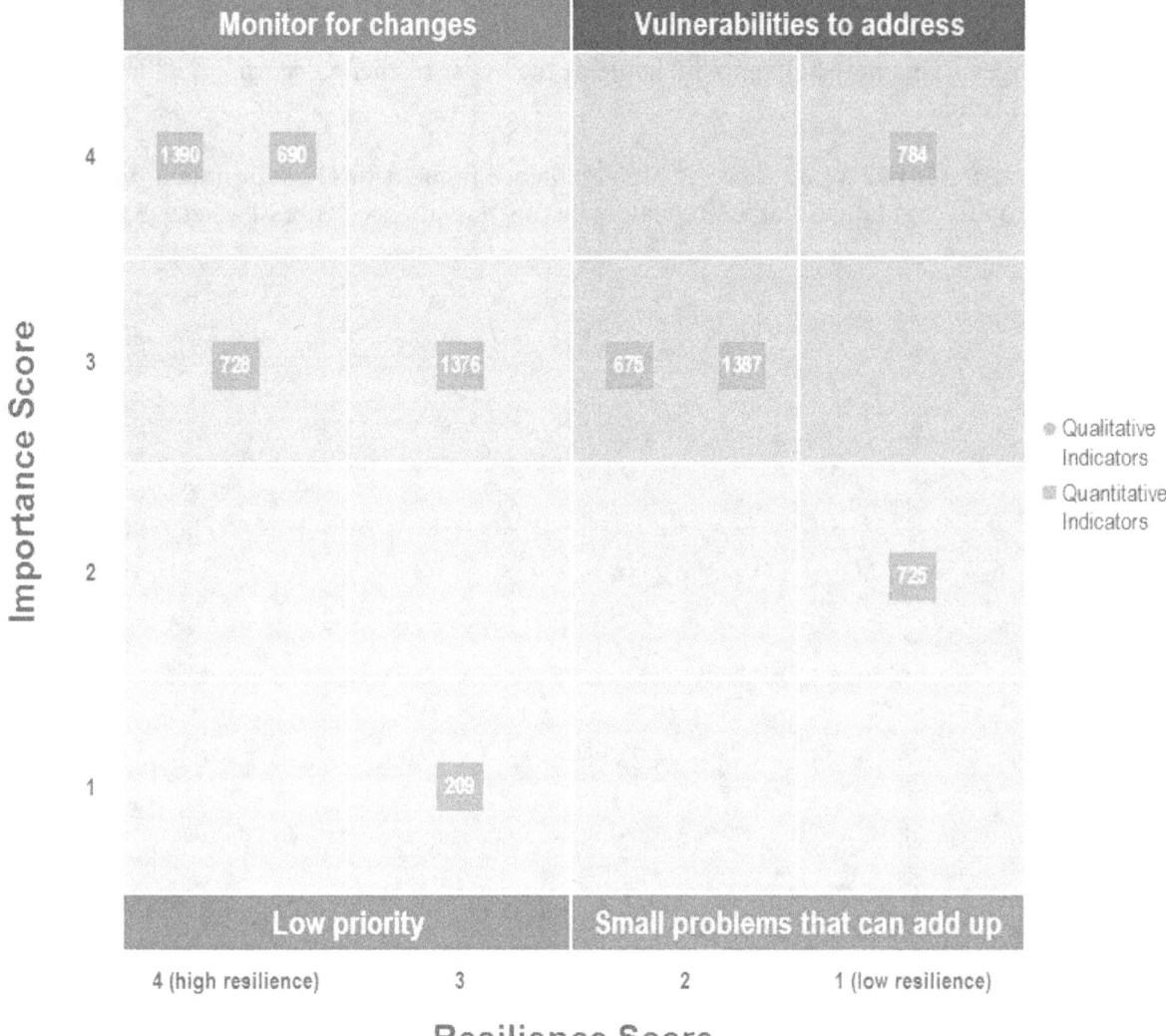

Figure 23. Worcester, MA people sector: Qualitative and quantitative indicator quadrant mapping.

As noted previously, data for the people sector are limited to quantitative indicator data. No responses were provided to the questions associated with this sector's qualitative indicators.

Planning resources for response to extreme events (handled by the Emergency Management Division of the Department of Public Health) are limited in availability and comprehensiveness. In addition, the capacity of emergency response systems and transportation resources and the capacity and distribution of public health works and emergency response resources in the event of an extreme event are somewhat to very limited. The number of police officers per capita is also low (0.0024). Transportation is a particular issue for vulnerable subpopulations. Response time is highly dependent on location, day of the week, and time of day. Maps showing data in space and at different days/times would be informative to gauge resilience. Fire response teams are routinely faster than medical emergency management services response teams because fire stations are spread out. Average fire response time was estimated as 3.0 minutes. The importance of response times depends on the situation and the nature of the emergency. For example, in the aftermath of a major storm with a limited number of ambulances, response times for large numbers of injured people would be critical compared to situations with large numbers of deaths.

However, climate change-related programs for adaptive behavior at the community level have been appropriately designed and promoted, although success has been limited, depending on the issue and the type of change being sought. For example, the city uses cooling centers (typically shopping malls) during extreme heat events. However, the elderly, especially those with asthma, are vulnerable because they cannot easily move to public cooling centers.

Urban planning and infectious disease response planning activities do account for the potential impacts of climate change and recognize potentially vulnerable subpopulations.

While availability of public health goods and services is only at risk if extreme climatic events or gradual climate change affect other city goods and services, loss of water and sanitation services can potentially create serious public health risks, especially for vulnerable subpopulations (18.3% of the population is vulnerable due to age).

Figure 23 shows that 78% of the indicators have an importance score of at least 3, with resilience scores distributed widely and relatively evenly across the resilience axis. Close to half of the indicators fall into the "monitor for changes" quadrant (high resilience/high importance) and 33% lie in the "vulnerabilities to address" quadrant.

E.2.1.6. *TELECOMMUNICATIONS*

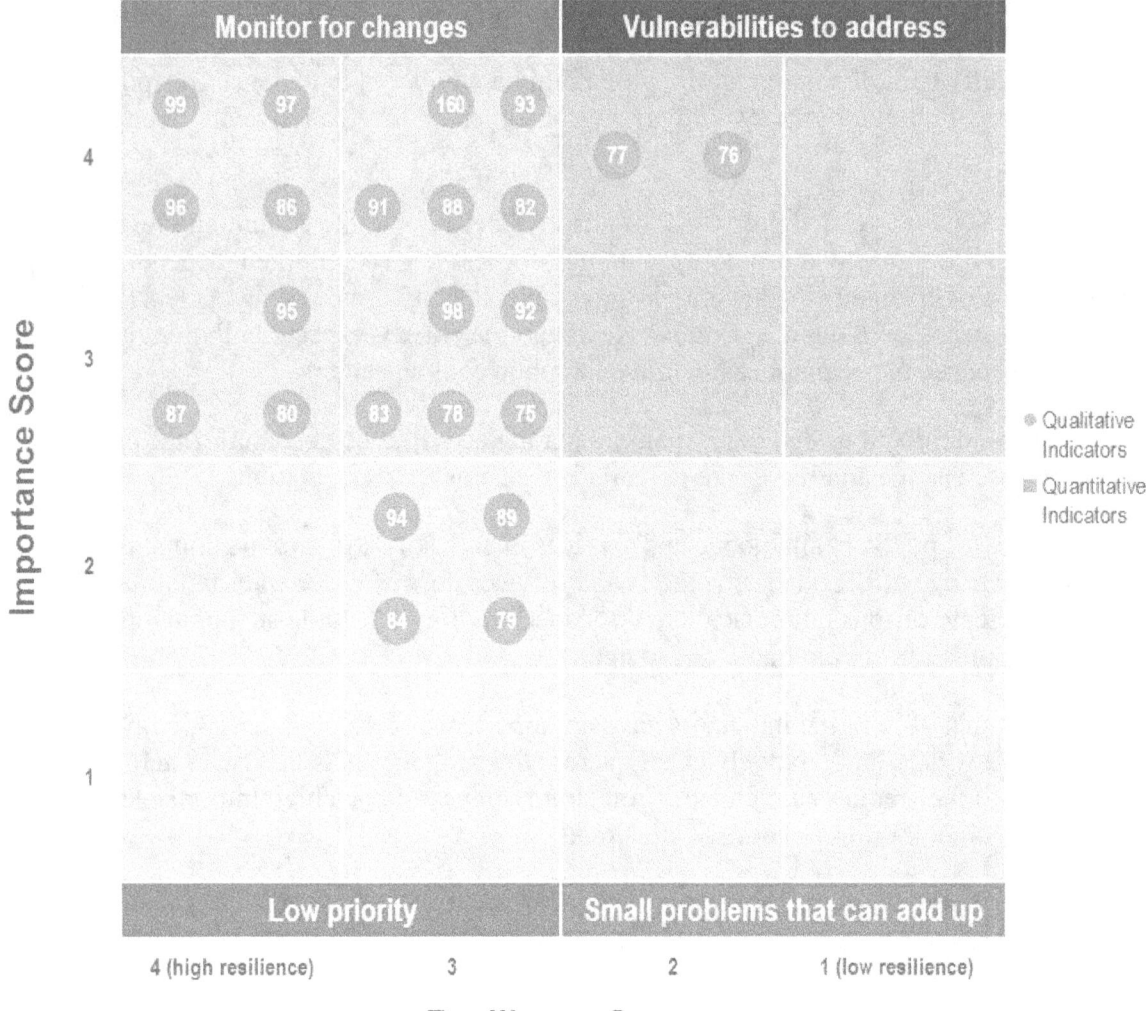

Figure 24. Worcester, MA telecommunications sector: Qualitative and quantitative indicator quadrant mapping.

The data gathered on the telecommunications sector came exclusively from qualitative indicators. No quantitative indicator data were available for relevant indicators.

Worcester's telecommunications sector generally demonstrates high resilience regarding emergency preparedness, robustness/vulnerability of the network and infrastructure, backup power and redundancy, and past experience. The city has experienced extreme weather and other similar events in recent years, and telecommunications services were only impacted to a limited extent.

Local authorities have established relationships with telecommunications service providers, and the systems and agreements in place ensure swift decision making and prioritization during emergencies. The city also has adequate communication systems in place to broadcast emergency information to the public.

Telecommunications infrastructure is generally not located where it would be vulnerable to high winds or flooding, and the infrastructure has the capacity for increased demand in emergency circumstances. It also unlikely that the capacity of first responder communication systems would be exceeded during a disaster or emergency, and the city has adequate backup telecommunications systems and power for those systems, as well as a great deal of telecommunications redundancy.

While disruption in the telecommunications sector may have impact the economy sector, there is only some risk that disruptions to other city goods and services (power, water, etc.) would impact the telecommunications sector.

However, the city appears to be less resilient in terms of a potential temporary loss of telecommunications and its impact on the local and regional economies, as well as the location of data centers, which are to some extent outside of the urban area (qualitative indicator #77). However, for all other relevant qualitative indicators, the participant indicated that the city's telecommunications sector is generally resilient.

Telecommunications resilience includes lines of communication between government and citizenry. In addition to 9-1-1 services, Worcester has an emergency notification system, ALERT Worcester, which contacts residents and businesses. The city's website provides information on preventing heat-related illness, the location and status of cooling centers during heat waves, and a citizen's guide to emergency preparedness. The website includes a voluntary emergency preparedness registry so that individuals with disabilities can provide information on their location and needs during emergencies (City of Worcester, 2013a).

As shown in Figure 24, the majority of the qualitative indicators (21 of 23) scored a 3 or above for resilience and 74% of the qualitative indicators (17 of 23) scored in the "monitor for changes" quadrant (high resilience/high importance), demonstrating that Worcester's telecommunications sector is resilient to climate change.

E.2.1.7. TRANSPORTATION

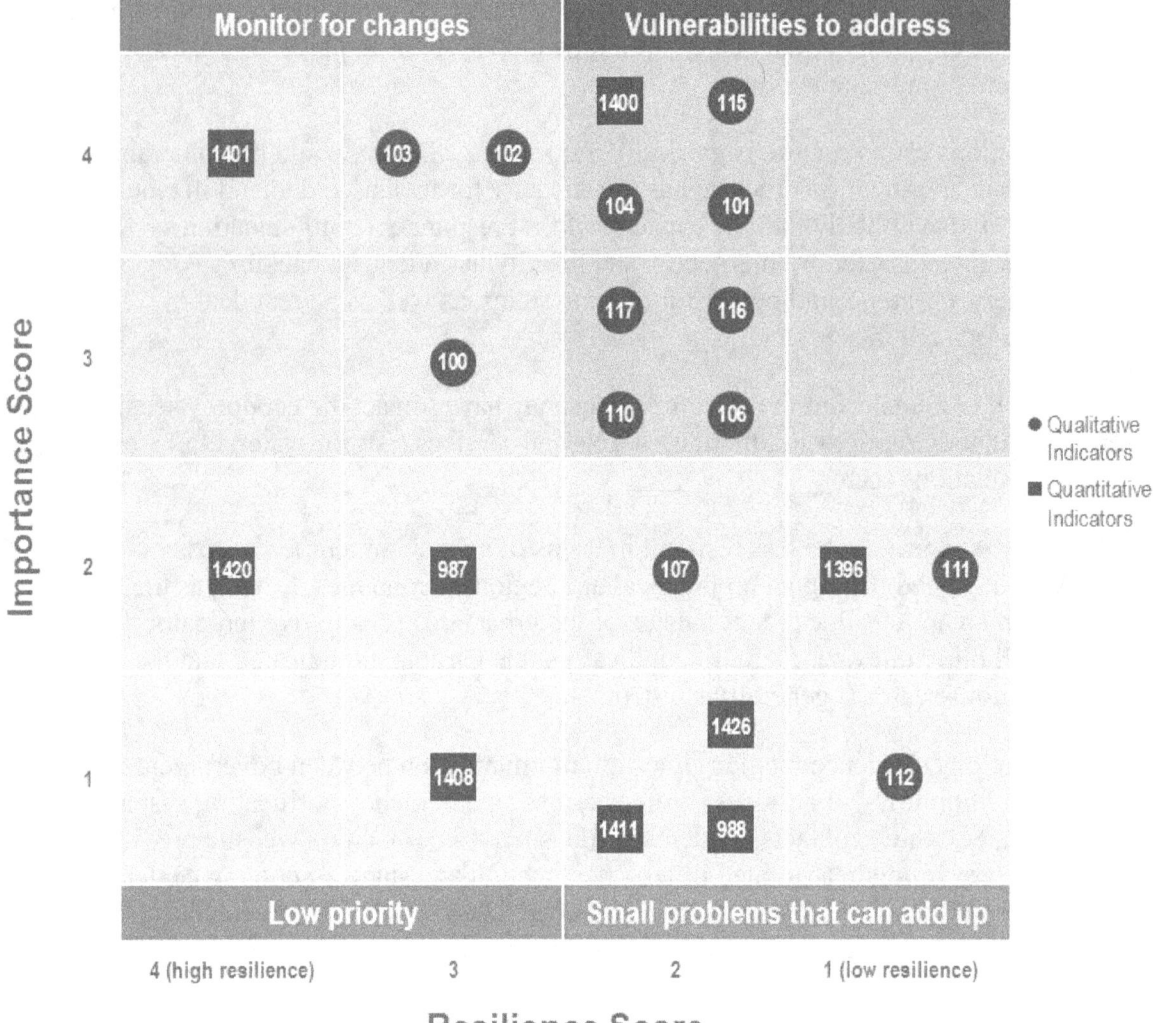

Figure 25. Worcester, MA transportation sector: Qualitative and quantitative indicator quadrant mapping.

Worcester's transportation is varied. The system includes a commuter rail line to and from Boston (owned by the Massachusetts Bay Transportation Authority (MBTA), also known as the "T"), Worcester Regional Transit Authority bus lines, and various local and state surface roads and highways for vehicle transit.

Worcester's transportation sector demonstrates limited resilience, which could have significant implications for the community's ability to respond to and recover from a major climatic event. It is also unclear whether the city could maintain adequate transportation services in the face of gradual effects on the sector due to climate change. For example, the length of time that would be required to restore major passenger rail transportation links in the urban area after a failure could be more than a week; however, the participant rated this question as only moderately important (score of 2).

The qualitative indicator scores did suggest that the city's transportation system is moderately resilient despite the lack of resilience or adaptation planning; that its redundancy is generally adequate; and that availability of transportation resources would not be heavily impacted if climate change or extreme climatic events affect other city goods and services. However, restoration of services in the event of a failure would be fairly slow. The length of time to restore major freight rail services in the event of a climate-related disruption depends on the nature and volume flow of the freight (food, medical goods, raw materials, etc.) and the nature of the disruption. The time to restore major high-traffic assets would depend on the nature of the specific asset and its disruption. For example, a November 2013 multiple-vehicle accident due to icy conditions on a section of interstate passing through downtown took a full day to clear. The participant noted that risk and recovery mapping of the transit and transportation system would be desirable for resilience planning.

The city also demonstrates low resilience related to community knowledge of evacuation procedures. While residents are slightly familiar with evacuation procedures within their own communities, coordination among neighboring communities and towns is at an early stage. In addition, residents are resistant to changing their preferred modes of transit unless there is a compelling reason or incentive. However, the Central Massachusetts Regional Planning Commission, in coordination with the Montachusett Regional Planning Commission, completed the first two phases of evacuation planning for Worcester in 2015. This included data gathering, mapping, and completion of a day-long tabletop exercise. Future plans include training local Emergency Management Directors on using the gathered data and identifying detours on major highways in the region.

Figure 25 reflects some, but not all, of these data. Data points vary widely on the importance axis, where approximately half of the qualitative and quantitative indicators have an importance score above 3 and half below 3. Over 65% of qualitative and quantitative indicators received a resilience score of 2 or lower, indicating that the transportation sector is relatively vulnerable. However, no scores fall in the lowest resilience (score of 1) and medium-to-high importance (scores of 3 and 4) quadrants, indicating that Worcester has taken some steps to address highly important factors related to transportation.

E.2.1.8. WATER

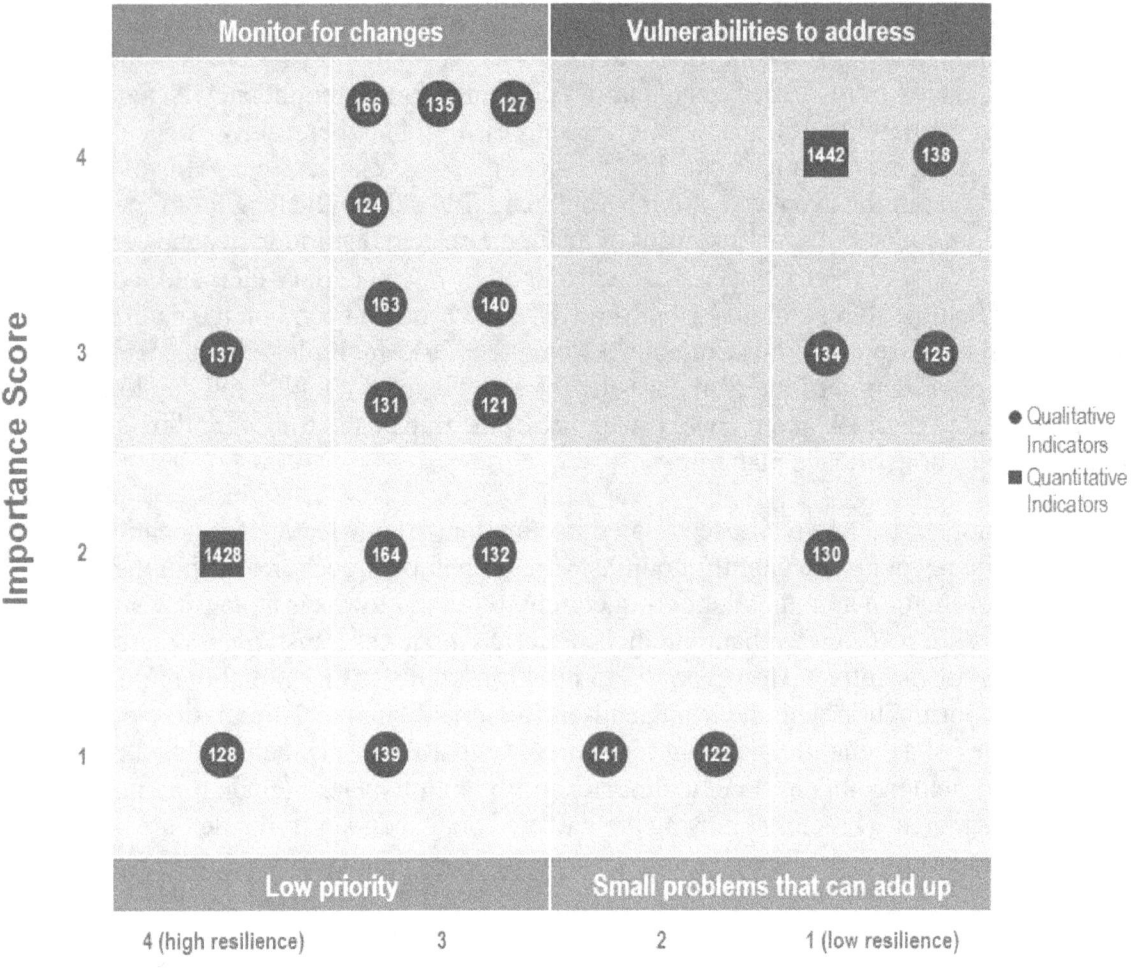

Figure 26. Worcester, MA water sector: Qualitative and quantitative indicator quadrant mapping.

The city's Department of Public Works and Parks oversees operation and maintenance of the city's drinking water infrastructure and supplies (10 surface water sources outside the city limits) and sewer infrastructure. The city's drinking water is treated at a water treatment plant at a rate of 50 million gallons per day, using a combination of ozone, coagulation, and filtration. The city has had no Safe Drinking Water Act violations in the past five years, but it received the lowest resilience rating for the ratio of water consumption to water availability.

Wastewater is treated at the Upper Blackstone WWTP before it is discharged into the Blackstone River. Since its construction, the District has completed over $170 million in improvements to the WWTP. In 2009, further improvements increased energy efficiency, provided solar power for the plant, and upgraded the solids management facilities. The improvements reduced the plant's carbon footprint and increased its treatment capacity (UBWPAD, 2013).

Further improvements to the drinking water and wastewater infrastructure to increase climate change resilience have been a challenge due to limited funding and environmental regulations set forth by federal and state legislation. In Worcester, the Millbury WWTP remains vulnerable to high-intensity storms in terms of both flooding (floodplain location proximate to the Blackstone River) and the limited capacity to handle high stormwater throughput. These vulnerabilities contributed to low resilience ratings.

In 2012, monitoring conducted by several federal agencies determined that most of central Massachusetts was in a moderate drought. Water consumption in the city peaked in 1988 around 27 million gallons per day and currently averages 22 million gallons per day. The Department of Public Works and Parks depends solely on user rates for revenue, requiring rate increases as consumption declines. The need for infrastructure improvements was highlighted by a November 2012 water main break that flooded parts of the Worcester State University campus, requiring water services to be shut off for the entire city.

Interconnectivity is a significant issue for this sector. The availability of water resources is at significant risk if other city services, particularly energy/power, are affected by climatic changes or events. Wastewater treatment typically has a high dependence on electrical power due to the energy-intensive unit processes of the primary, secondary, and tertiary treatment stages. There is full backup power on the collection side, but only some on the treatment side.

Despite concerns about the limitations of backup power, the impact loss of power has on the water sector, and the lack of hierarchy of water uses during a shortage or emergency, the city otherwise demonstrates resilience with respect to emergency preparedness, emergency response, and redundancy. The water system has emergency connections with adjacent water systems or emergency sources of supply, as well as redundant treatment and distribution systems. The drinking water treatment plant also has redundant treatment chemical supplies. In addition, a WARN provides technical resources and support during emergencies. Worcester is also part of a regional stormwater initiative with neighboring towns.

The city has also undertaken water-related planning efforts, incorporating past experiences into planning approaches for water shortages or increases in the frequency of overflows. The city also has programs related to long-term maintenance of water supplies, has inventoried storm sewers and drains to storm sewers, and has used these inventories in planning efforts. However, customer familiarity with conservations measures and the measures' implementation is somewhat limited. In general, the city's properties are not equipped to harvest rainwater or recharge groundwater, and residents practice rainwater harvesting on a very limited scale. Lawn watering habits are the primary concern with regard to water conservation.

Figure 26 shows that the data from the water sector spreads fully across both the importance and resilience axes. Based on the available data, 71% of the qualitative and quantitative indicators have resilience scores of 3 or above, demonstrating that Worcester's water sector is relatively resilient. The three topics located in the "vulnerabilities to address" quadrant (low resilience/high importance), which may be of greatest concern to the city, include the city's lack of hierarchy of water uses during a shortage or emergency; the fact that properties in the city are not equipped to harvest rainwater; and the fact that the availability of water resources is at significant risk if other city services, particularly energy/power, are affected by climatic changes or events.

E.2.2. SUMMARY OF WORCESTER, MA FINDINGS

As with Washington, DC, the findings for Worcester indicate mixed resilience. In addition, the comprehensiveness of the results is limited by lack of data (quantitative indicator data or qualitative responses to questions) for the people, energy, and telecommunications sectors. The City demonstrated relatively high resilience in respect to the economy, telecommunications, water, and natural environment sectors. Resilience in the remaining sectors is largely mixed, although more significant potential vulnerabilities were identified in the energy sector. However, as no quantitative indicator data were available for the energy sector, a more complete picture of the city's vulnerability was not available.

Positive trends in economic diversification point to the potential for additional resources for and interest in future adaptation activities (as well as the need for such activities). However, data collected for the land use/land cover sector indicate potential vulnerabilities in the economy sector, due to the location of key infrastructure and continued development (without concern for retrofitting) in areas that are vulnerable to extreme events. Disruption to water services could have significant impacts on other sectors, and vice versa. However, the city otherwise demonstrates resilience with respect to emergency preparedness and response, as well as redundancy within both the water and telecommunications sectors.

While the framework was tested and implemented through a workshop process in Washington, DC, the interviews in Worcester were conducted primarily with one sector representative, as the framework design intended. Therefore, the limitations observed in Worcester related to data availability; previous or ongoing efforts related to climate change adaptation and resilience; and resources may be more indicative of the challenges to implementing the framework and addressing vulnerabilities that like-sized urban communities face.

APPENDIX F. COMPARISON OF RESULTS FOR WASHINGTON, DC AND WORCESTER, MA

This appendix presents cross-city comparison visualizations. While the value of this visualization is limited by the small sample size of two cities, the value will increase as the tool is used for evaluating the resilience of additional cities because patterns will become clearer as additional data are gathered and confidence in those patterns increases.

F.1. City Comparison

Washington, DC and Worcester, MA provide contrasting examples of the risks faced by urban areas and the resources that mid- to large-sized communities may have to plan for climate-resilient futures. Table 14 highlights some of the key features of both cities. Choosing these contrasting cities allows cities within a broad spectrum in terms of resources, planning, and risk to understand the applications and potential outcomes of using the tool. It also allows us to test the strengths and weaknesses of the tool's methodology in a wide range of conditions and provides preliminary insight into the range (or potential lack) of risk exposures across cities with different geographic, economic, population, and historical characteristics.

Table 14. Washington, DC and Worcester, MA metrics at a glance

	Washington, DC	Worcester, MA	USA average
Population	658,893 (U.S. Census Bureau, 2013a)	183,016 (U.S. Census Bureau, 2013c)	
Population growth, 2010–2013	+ 7.9% (U.S. Census Bureau, 2013a)	+ 1.1% (U.S. Census Bureau, 2013c)	+ 2.5% (U.S. Census Bureau, 2013a)
Median household income	$65,830 (U.S. Census Bureau, 2013a)	$45,932 (U.S. Census Bureau, 2013c)	$53,046 (U.S. Census Bureau, 2013a)
Percentage below poverty level	18.6% (U.S. Census Bureau, 2013a)	21.4% (U.S. Census Bureau, 2013c)	15.4% (U.S. Census Bureau, 2013a)
Total number of firms	55,887 (U.S. Census Bureau, 2013a)	11,799 (U.S. Census Bureau, 2013c)	
Chief industries	Federal services, tourism (U.S. Census Bureau, 2013b)	Education, medical, biotech (City of Worcester, 2004; Research Bureau, 2008)	
Topography	Coastal plain	Hilly	
Region	Southeastern Seaboard	New England	
Hazards	Sea level rise, hurricanes, drought, heat waves, severe storms (MWCOG, 2013a)	Drought, heat waves, tornados, severe storms, blizzards (CMRPC, 2012)	
Climate adaptation planning	High	Low	
Capacity for climate adaptation	High	Low	
Similar cities	New York City, Boston, Atlanta, Miami	Cleveland, Pittsburgh, Detroit, Buffalo, St. Louis, Providence	

F.2. Results—quadrant map comparisons

Figure 27 maps the qualitative and quantitative indicator sector averages from both Washington, DC and Worcester, MA on a quadrant graph. This approach does not show intrasector areas of higher or lower vulnerability, but it facilitates comparison between the two cities and between qualitative and quantitative indicator-based results. Overall, the results for both cities for all sectors cluster moderately tightly, with the center of the cluster falling in the "vulnerabilities to address" quadrant. There does not appear to be a close match of data points representing the same sectors between cities. Given the low sample size and the lack of spread in the data, there is an insufficient basis to conclude that any sector is more or less vulnerable than another overall.

For the qualitative indicators, the spread of the data is slightly less than one point in both resilience and importance scores. The spread for the quantitative indicator data is greater, but it is difficult to determine whether this spread is meaningful because much less quantitative indicator data were available—particularly for Worcester. The narrow range of variability in the results for the two cities is striking, given the differences in indicator data quality and availability between the two locations.

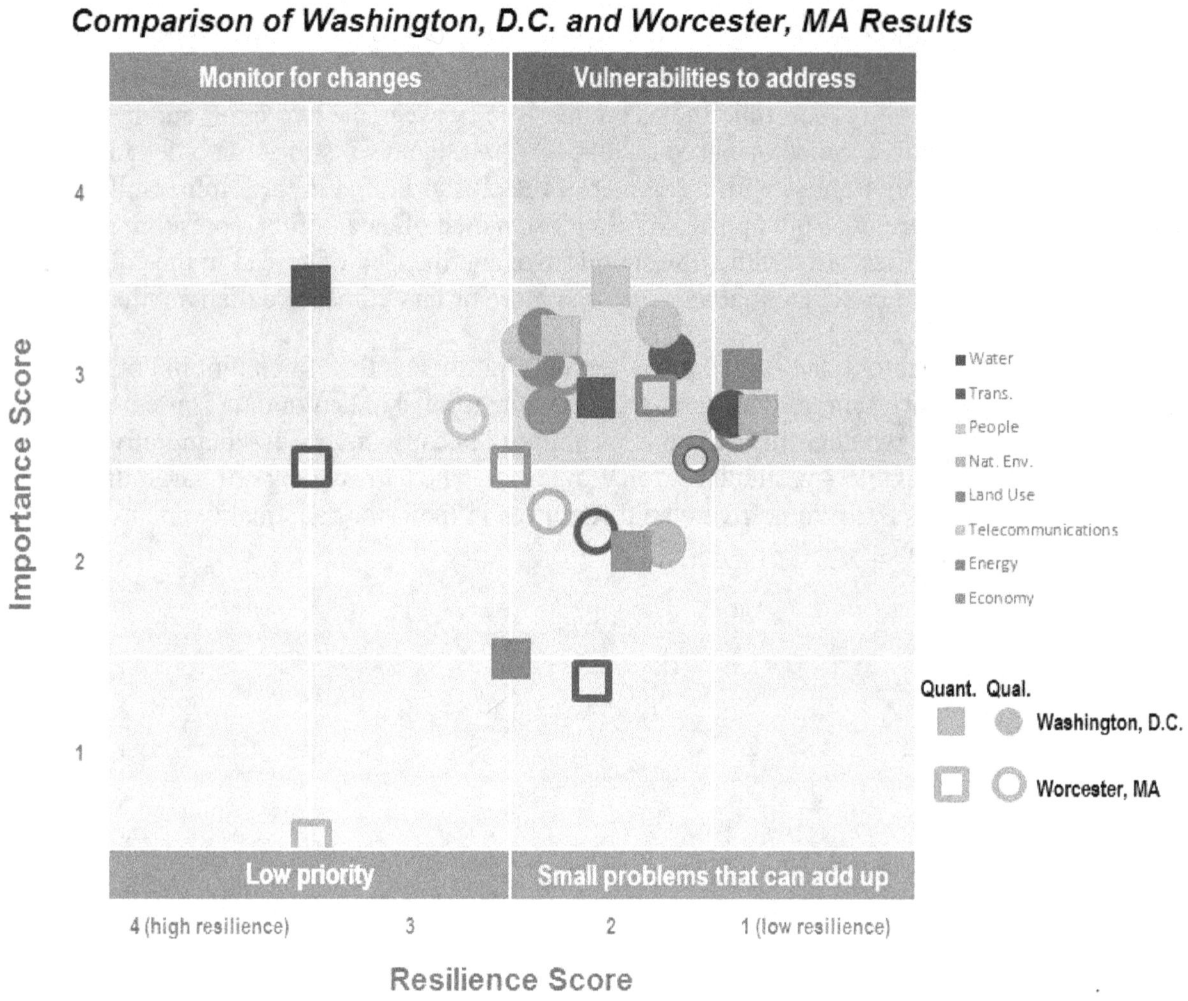

Figure 27. Washington, DC and Worcester, MA: Average quantitative indicator and qualitative indicator score quadrant mapping.

APPENDIX G. QUALITATIVE INDICATORS: ORDERED

Qualitative Indicator ID#	Sector	Name/Question
1	Economy	Is the economy of the urban area largely independent, or is it largely dependent on economic activity in other urban areas?
2	Economy	Does the urban area have mechanisms to help businesses quickly return to normal operations?
3	Economy	If jobs are lost in one sector of the urban area, does the capacity exist to expand the economy and job opportunities in another sector?
4	Economy	Has the vulnerability of critical infrastructure been assessed? Are there plans to relocate or protect vulnerable infrastructure in ways that promote resilience and protect other infrastructure and properties?
5	Economy	Has the urban area's resilience to major changes in energy policy/prices been assessed?
6	Economy	Is funding available for adaptive development projects that could also serve as recreation areas (e.g., retention areas along waterways that could also serve as parks)? Are such multipurpose projects required or are there incentives for these projects?
7	Economy	Is a significant portion of the population of the urban area either seasonal residents or transient populations that may have a lesser degree of understanding of changes occurring within that area?
8	Economy	How many people are in place to respond to emergencies, and what is the level of communication connectivity of emergency response teams and offices?
9	Economy	Is comprehensive adaptation planning possible with the urban area's current resources? If so, is adaptation planning already occurring?
10	Economy	Is planning for climate change adaptation in the urban area incorporated into one office within the local government, or is planning spread out across several offices within the government?
11	Economy	How flexible are planning processes for short-term and long-term responses? For example, is there flexibility in changing planning priorities if necessary?
12	Economy	Does adaptation planning for the urban area include retrospective analyses of past events (including analyses of past climate events in other cities if helpful) to help determine whether decisions on adaptation measures would be effective?
13	Economy	Does adaptation planning for the urban area consider the costs and benefits of possible decisions, and does it encourage both pre- and post-event evaluations of the effectiveness of adaptation measures?

14	Economy	Do adaptation plans account for tradeoffs between the less resilient but lower cost strategy of increasing protection from climatic changes and the more resilient but higher cost strategy of moving residents from the most vulnerable portions of the urban area? (One example of such a tradeoff is in coastal cities, where some areas can be protected by a seawall, or households and institutions in vulnerable areas can be moved inland. Do current adaptation plans account for the resilience-cost tradeoffs in this decision?)
15	Energy	Do you have a diverse energy portfolio?
16	Energy	Are there redundant systems in place for coping with extreme events?
17	Energy	To what extent do energy supplies come from outside the metropolitan area?
18	Energy	Is the availability of energy goods and services at risk if other city goods and services (e.g., water, transportation, telecommunications) are affected by extreme climatic events or gradual climatic changes?
19	Energy	How many minutes per year or hours per year do you have power outages?
20	Energy	What is the response time to restore electrical power after an outage?
21	Energy	Does capacity exist to handle a higher peak demand or peaks at different times?
22	Energy	To what extent have efforts been made to reduce energy demand?
23	Energy	What are the opportunities for distributed generation sources (e.g., different capacity for energy generation from different sources including renewable)?
24	Energy	Are there smart grid opportunities to manage demand?
25	Land Use/Land Cover	Can resilience planning/adaptation be incorporated into existing programs that communities engage in regularly (e.g., zoning, hazard mitigation plans)?
26	Land Use/Land Cover	Has the city made efforts to use urban forms to mitigate climate change impacts and to maximize benefits (e.g., urban tree canopy cover)?
27	Land Use/Land Cover	Are urban forms used that address (lessen) urban heat island effects (e.g., through increasing evapotranspiration or increasing urban ventilation)?
28	Land Use/Land Cover	Does zoning encourage green roofs or other practices that reduce urban heat?
29	Land Use/Land Cover	Are there mechanisms to support tree shading programs in urban areas (to reduce urban heat and improve air quality)? Are there innovative ways to fund such programs?
30	Land Use/Land Cover	Have land use/land cover types, such as soil and vegetation types and areas of tree canopy cover, been inventoried, and are these inventories used in planning?
31	Land Use/Land Cover	What percentage of open/green space is required for new development (to encourage increases in such space)?

32	Land Use/Land Cover	Are there mechanisms for the local government to purchase land that is unfavorable for redevelopment due to the results of extreme events (e.g., flooding from a hurricane)? If so, what are those mechanisms?
33	Land Use/Land Cover	Are there policies or zoning practices in place that will allow transfer of ownership of undeveloped land subject to flooding or excessive erosion to the city (or allow nonpermanent structures only)? Are these policies or zoning practices enforced?
34	Land Use/Land Cover	Where developed land is located in areas vulnerable to extreme events, are resilient retrofits being implemented or planned?
35	Land Use/Land Cover	Are there codes to prevent development in flood-prone areas?
36	Land Use/Land Cover	Are regulations in place regarding whether communities that are affected by floods will be rebuilt in the same location?
37	Land Use/Land Cover	Have the regulations regarding rebuilding of communities affected by floods been enforced to date?
38	Land Use/Land Cover	Do incentives exist to integrate green stormwater infrastructure into infrastructure planning to mitigate flooding?
39	Land Use/Land Cover	Are there incentives to reduce the amount of impervious surface, to prevent development in floodplains, to use urban forestry to reduce impacts, to use green infrastructure for stormwater management, etc.?
40	Land Use/Land Cover	To what extent was green infrastructure selected to provide the maximum ecological benefits?
41	Land Use/Land Cover	Has green infrastructure maintenance been built into the budget?
42	Natural Environment	Is the availability of environmental/ecosystem goods and services at risk if other city goods and services (e.g., power, water, telecommunications) are affected by extreme climatic events or gradual climatic changes?
43	Natural Environment	What regulatory and planning tools related to air quality, water quality, and land use are already available locally? For example, does the urban area have invasive plant ordinances or tree planting requirements?
44	Natural Environment	Do plans exist for increasing open and green space?
45	Natural Environment	Has the continuity of open or green spaces been assessed and addressed in planning efforts?
46	Natural Environment	Do native plant or animal species lists exist for the urban area, and are these species (rather than nonnative species) used in green infrastructure?
47	Natural Environment	Does the urban area coordinate with other nearby entities on water quality?
48	Natural Environment	To what degree do local versus distant sources influence air quality?
49	Natural Environment	Does the urban area have air quality districts?

#	Category	Question
50	Natural Environment	Has an air quality analysis been completed at multiple scales/resolutions?
51	Natural Environment	Does the urban area have health warnings or alerts for days when air quality may be hazardous?
52	Natural Environment	Has an analysis of areas with good ventilation (e.g., aligned with prevailing breezes, good tree canopy cover) been completed?
53	Natural Environment	Do plans exist for preserving areas with good ventilation (e.g., those aligned with prevailing breezes)?
54	Natural Environment	Does the urban area have a district-scale (i.e., higher resolution than city scale) thermal comfort index?
55	People	How available and how comprehensive are your planning resources for responding to extreme events?
56	People	Are government-led, community-based, or other organizations actively promoting adaptive behaviors at the neighborhood or city level?
57	People	Do policies and outreach/education programs promote behavioral changes that facilitate climate change adaptation?
58	People	Are emergency response staff well trained to respond to large-scale extreme weather events?
59	People	Is the distribution of public health workers and emergency response resources appropriate for the population that would be affected during an extreme event?
60	People	Is there sufficient capacity in public health and emergency response systems for responding to extreme events?
61	People	Does the city have the capacity to provide public transportation for emergency evacuations?
62	People	What evacuation and shelter-in-place options are available to residents in the event of a heat wave?
63	People	Do plans exist to provide public access to cooling centers or for other heat adaptation strategies (e.g., opening public swimming pools earlier or later than normal, using fire hydrants for cooling), given predicted climatic changes?
64	People	Is the health care community, including primary care physicians, prepared for changes in patients' treatments necessitated by climate change (e.g., emerging infectious diseases)?
65	People	Is the availability of public health goods and services at risk if other city goods and services (e.g., power, water, public transportation) are affected by extreme climatic events or gradual climatic changes?
66	People	Do public health programs incorporate longer time frames (e.g., 10 or more years), and do they address climate change-related health issues (e.g., movement of deer ticks to more northerly locations)?
67	People	Have public health agencies identified infectious diseases and/or disease vectors that may become more prevalent in the urban area under the expected climatic changes?

#	Category	Question
68	People	Have public health agencies developed plans for responding to increased disease and vector exposure in ways that may reduce the associated morbidity/mortality?
69	People	Do planners in the urban area know the demographic characteristics of populations vulnerable to climate change?
70	People	Do planners in the urban area know the locations of populations most vulnerable to climate change effects?
71	People	Are there services and emergency responses aimed at quickly reaching vulnerable populations during power outages?
72	People	Are policies and programs to promote adaptive behavior designed with frames/messaging that reach the critical audiences in the urban area?
73	People	Are policies and programs to promote adaptive behavior designed and implemented in ways that promote the health and well-being of vulnerable populations?
74	People	Are policies and programs to promote adaptive behavior evaluated in ways that take into account vulnerable populations?
75	Telecommunications	What natural disasters has the area experienced in the past, and what services were retained or largely unaffected despite these disasters?
76	Telecommunications	How would a temporary loss of telecommunication infrastructure affect the local and regional economies?
77	Telecommunications	Are data centers located within or outside of the urban area?
78	Telecommunications	For each telecommunication service, are there key nodes whose failure would severely affect the service?
79	Telecommunications	How robust is the telecommunication network in terms of resilience to damage to or failure of key nodes?
80	Telecommunications	Are there parts of the telecommunication infrastructure that are particularly vulnerable to high temperatures or prolonged high temperatures?
81	Telecommunications	Are there satellite-based communications on frequency bands (e.g., the Ka band) that are vulnerable to wet-weather disruption?
82	Telecommunications	Are your telecommunication infrastructure components located wisely with respect to your anticipated climate stressors (i.e., aboveground, underground, or serviced by satellite)?
83	Telecommunications	Are aboveground infrastructure components vulnerable to wind (e.g., cell towers)?
84	Telecommunications	Are belowground infrastructure components vulnerable to rising water or salt water intrusion?
85	Telecommunications	If the area has satellite-based communications that are vulnerable to wet-weather disruption, does the area have a backup tower network?

86	Telecommunications	Does your community have sufficient access to backup telecommunication systems? What is the capacity of the telecommunication infrastructure?
87	Telecommunications	Is backup power for telecommunication systems provided? If so, is it provided by diesel generators?
88	Telecommunications	What is the extent of telecommunication redundancy? Do first responders and the public have multiple communication options, served by different infrastructures?
89	Telecommunications	What percentage of telecommunication system capacity is required for the baseline level of use?
90	Telecommunications	Does telecommunication infrastructure have the capacity for increased public demand in an emergency?
91	Telecommunications	Do local authorities have established relations with telecommunication infrastructure service providers? Are emergency protocols and plans in place?
92	Telecommunications	Do local private-sector telecommunication infrastructure service providers have the authority and resources to make quick decisions and implement them in and after an emergency?
93	Telecommunications	Can local authorities and telecommunication providers give first-responder and decision-maker communications priority during an expected surge in traffic in emergency situations?
94	Telecommunications	Are public-address systems (e.g., loudspeakers, text messages, radio broadcasts, emergency television broadcasts) in place to provide instructions to the public in case of an emergency?
95	Telecommunications	What modes do authorities in the urban area use to communicate emergency information and alerts? Are these modes low or high bandwidth?
96	Telecommunications	What is the likelihood that the capacity of local first responder communication systems would be exceeded during a disaster?
97	Telecommunications	Does the area have access to backup emergency call/response (911) networks if the primary networks fail or are overloaded?
98	Telecommunications	Is the availability of telecommunication goods and services at risk if other city goods and services (e.g., power, water, transportation) are affected by extreme climatic events or gradual climatic changes?
99	Telecommunications	Do telecommunication systems have enough energy and water supply to handle an extra load in the case of sudden natural disasters?
100	Transportation	Is the availability of transportation goods and services at risk if other city goods and services (e.g., power, water, telecommunications) are affected by extreme climatic events or gradual climatic changes?
101	Transportation	How much risk is assumed in the design of transportation systems (bridges, culverts), and does it span the anticipated changes in precipitation, temperature, and storm intensities under climate change?

102	Transportation	How resistant to potential impacts of climate change are critical transportation facilities (e.g., high-traffic vehicle or rail bridges, tunnels)?
103	Transportation	What degree of redundancy exists for major transportation links? Are there single points of failure? What are the implications of losing a particular link, and how rapidly can you recover?
104	Transportation	What length of time would be required to restore major high-traffic vehicle transportation facilities in the urban area if they experience a failure?
105	Transportation	Are any portions of the transportation system less important if the duration of the disturbance is a few days? What if the duration of the disturbance is more on the order of weeks?
106	Transportation	To what extent is the area dependent on long-range transportation of goods and services versus locally available goods and services (food, energy, etc.)?
107	Transportation	What flexibility has been built into the transportation system (different modes)?
108	People	How accessible are different modes of transportation (e.g., to what proportion of the population, what subpopulations [vulnerable people])?
109	People	What proportion of the population has limited access to transportation options due to compromised health or lower income levels? For what proportion of this population might transportation failures be life-threatening (e.g., due to reduced access to specialized medical care or equipment)?
110	Transportation	How familiar is the community with evacuation procedures?
111	Transportation	What length of time would be required to restore major passenger rail transportation facilities in the urban area if they experience a failure?
112	Transportation	What length of time would be required to restore major freight rail transportation facilities in the urban area if they experience a failure?
113	Transportation	What length of time would be required to restore major bicycle and pedestrian transportation links in the urban area if they experience a failure?
114	Transportation	Are urban areas set up to provide accessibility (e.g., to jobs) if mobility is interrupted or impeded?
115	Transportation	Do current planning regimes include proactive resilience building, or is only reactive disaster response being addressed?
116	Transportation	Are there funding mechanisms that exist or could be put into place to complete the necessary work on the transportation system to adapt to anticipated climatic changes and increased risks?
117	Transportation	Do plans exist to replace aging infrastructure? If so, do these plans account for the anticipated impacts of climate change on this infrastructure?
118	Transportation	Are the materials currently in use in transportation systems, such as the common asphalt formulations and rail types, compatible with anticipated changes in temperature?

119	Transportation	Have new or innovative materials been tested that may be more capable of withstanding the anticipated impacts of climate change (e.g., higher temperatures)?
120	Transportation	To what extent is green infrastructure implemented or planned to reduce climate change impacts on transportation systems?
121	Water	Does the water supply draw from a diversity of sources?
122	Water	To what extent do water supplies come from outside the metropolitan area?
123	Water	Is there a recharge plan in place for groundwater supplies?
124	Water	Do programs for long-term maintenance of water supplies (e.g., erosion control methods, reforestation of the watershed) exist?
125	Water	Is there a hierarchy of water uses to be implemented during a shortage or emergency?
126	Water	Does the water system have emergency interconnections with adjacent water systems or other emergency sources of supply?
127	Water	Are water and wastewater treatment plants located in a flood zone?
128	Water	Are groundwater supplies susceptible to salt water intrusion and sea level rise?
129	Water	If groundwater supplies are susceptible to salt water intrusion and sea level rise, is the water treatment plant equipped to deal with higher levels of salinity?
130	Water	Does treatment capacity exist to accommodate nutrient loading?
131	Water	Does the drinking water treatment plant have redundant treatment chemical suppliers?
132	Water	Are there redundant drinking water systems in place for coping with extreme events, including supply, treatment, and distribution systems?
133	Water	Is backup power for water supply, treatment, and distribution systems provided?
134	Water	How diverse are individual properties (i.e., are they equipped to harvest rainwater or recharge groundwater so they can create or augment local water supplies)?
135	Water	Are there redundant wastewater and stormwater systems in place for coping with extreme events, including collection systems and wastewater treatment systems?
136	Water	Does a water/wastewater agency response network provide technical resources/support to the urban area's water system during emergencies?
137	Water	Have storm sewers and drains to storm sewers been inventoried, and are these inventories used in planning?

138	Water	Is the availability of water goods and services at risk if other city goods and services (e.g., power, transportation, public health) are affected by extreme climatic events or gradual climatic changes?
139	Water	Has the water utility conducted a water audit to identify current losses (e.g., leaks, billing errors, inaccurate meters, unauthorized usage)?
140	Water	To what extent have efforts been made to reduce water demand?
141	Water	Are customers familiar with water conservation measures, and are they willing to implement these measures?
142	Land Use/Land Cover	Are coastal hazard maps with 1-meter altitude contours available, and are these maps used in planning?
143	People	Are early warning systems for meteorological extreme events available?
147	Energy	Do municipal managers draw on past data/experiences of extreme weather events to assess the effects of these events on oil and gas availability and pricing? (DOE, 2013)
148	Energy	Has the city consulted with local power companies to develop plans for potential increases in electricity demand for summer cooling? (DOE, 2013)
149	Energy	Has the city coordinated with local water suppliers and power generation facilities to discuss potential climate-induced water shortages and their impacts on cooling the power generation facilities? (DOE, 2013)
150	Energy	Do municipal managers in coastal areas consider the impacts of sea level rise on power generation facilities?
151	Land Use/Land Cover	Have institutional land practices (i.e., zoning, land use planning) potentially been hindered by other government agencies seeking to shift financial resources when it comes to climate change planning?
152	Land Use/Land Cover	Does knowledge of historical land use/land cover changes contribute to planners' understanding of climate stresses?
153	Land Use/Land Cover	Have specific historical land use/land cover changes been recognized as increasing or decreasing vulnerability to climate stresses?
154	Land Use/Land Cover	Does the city consider the knowledge of local academic research and other stakeholders (e.g., farmers, forest managers, land use managers) in land use planning related to climate resilience?
158	People	Do municipal managers consider local stakeholder knowledge and local resources (e.g., libraries, archives) in climate change resilience planning?
160	Telecommunications	Have city planners consulted with other city governments with similar telecommunication systems to learn from their experience with natural disasters and prepare for similar events?
162	Transportation	Have municipalities considered new methods of designing roads/bridges to prepare for heavily traveled routes during an extreme climate event (e.g., coastal evacuation routes)?
163	Water	Have water utility companies incorporated past experience or experience from other locations/utilities into developing plans for water shortages related to climate-induced stresses?

164	Water	Does the water department or utility for the city consider past experience in addressing anticipated increases in the frequency of sewer overflows?
165	Economy	What financial capacity is indicated by the city's bond ratings?
166	Water	Is backup power for wastewater collection and treatment provided?
167	Land Use/Land Cover	In general, what is the monetary value of infrastructure located within the 500-year floodplain in the city?
168	Transportation	How resistant to potential impacts of climate change are critical nonroad transportation facilities (e.g., high-traffic rail bridges, tunnels)?
169	Transportation	Do plans exist to replace aging infrastructure? If so, do these plans account for the anticipated impacts of climate change on this infrastructure?

APPENDIX H. QUANTITATIVE INDICATORS: ORDERED

Quantitative Indicator ID#	Sector	Name	Definition
17	Natural Environment	Altered wetlands (percentage of wetlands lost)	This indicator reflects the percentage of wetland areas that have been excavated, impounded, diked, partially drained, or farmed.
51	Land Use/Land Cover	Coastal Vulnerability Index rank	This indicator reflects the Coastal Vulnerability Index rank. The ranks are as follows: 1 = none, 2 = low, 3 = moderate, 4 = high, 5 = very high. The index allows six physical variables to be related in a quantifiable manner that expresses the relative vulnerability of the coast to physical changes due to sea level rise. The six variables are: a = geomorphology; b = coastal slope (%); c = relative sea level change (mm/year); d = shoreline erosion/accretion (m/year); e = mean tide average (m); e = mean wave height (m).
66	Natural Environment	Percentage change in disruptive species	This indicator reflects the percentage change in disruptive species found in metropolitan areas. Disruptive species are those that have negative effects on natural areas and native species or cause damage to people and property.
194	Land Use/Land Cover	Percentage of natural area that is in small natural patches	This indicator measures the percentage of the total natural area in a city that is in patches of less than 10 acres. Smaller patches of natural habitat generally provide lower quality habitat for plants and animals and provide less solitude and fewer recreational opportunities for people. About half of all natural lands in urban and suburban areas are in patches smaller than 10 acres.
209	People	Percentage of population living within the 500-year floodplain	This indicator reflects age of population living within the 500-year floodplain.
254	Land Use/Land Cover	Ratio of perimeter to area of natural patches	This indicator is calculated as the average ratio of the perimeter to area.
273	Natural Environment	Percentage of total wildlife species of greatest conservation need	This indicator reflects the percentage of total wildlife species that are listed as having the "greatest conservation need."

284	Natural Environment	Physical Habitat Index (PHI)	PHI includes eight characteristics (riffle quality, stream bank stability, quantity of woody debris, in-stream habitat for fish, suitability of streambed surface materials for macroinvertebrates, shading, distance to nearest road, and embeddedness of substrates). Scores range from 0–100 (81–100 = minimally degraded, 66–80 = partially degraded, 51–65 = degraded, 0–50 = severely degraded).
308	Land Use/Land Cover	Percentage of city land that is urban and suburban	This indicator presents the extent/acreage of urban and suburban areas (i.e., not rural) as a percentage of the total land area
322	People	Percentage of population affected by waterborne diseases	This indicator reports the percentage of population affected by waterborne diseases.
326	Natural Environment	Wetland species at risk (number of species)	Number of wetland and freshwater species at risk (rare, threatened, or endangered).
393	People	Percentage of vulnerable population that is homeless	This indicator reflects the percentage of population 65 and older and under 5 years that is homeless.
437	Water	Percentage change in streamflow divided by percentage change in precipitation	This indicator reflects percentage change in streamflow (Q) divided by percentage change in precipitation (P) for 1,291 gauged watersheds across the continental U.S. from 1931 to 1988.
460	Natural Environment	Macroinvertebrate Index of Biotic Condition	The Benthic Index of Biotic Integrity (BIBI) score is the average of the score of 10 individual metrics, including Total Taxa Richness, Ephemeroptera Taxa Richness, Plecoptera Taxa Richness, Trichoptera Taxa Richness, Intolerant Taxa Richness, Clinger Taxa Richness and Percentage, Long-Lived Taxa Richness, Percentage Tolerant, Percentage Predator, and Percentage Dominance.
465	Natural Environment	Change in plant species diversity from pre-European settlement	Change in the plant species diversity from pre-European settlement (baseline) to present, within a given city/area.
675	People	Asthma prevalence (percentage of population affected by asthma)	This indicator presents asthma prevalence for U.S. children (age 0–17) and adults (age 18 and older). It is calculated as the percentage of population reporting asthma. Asthma attack prevalence is based on the number of adults/children who reported an asthma episode or attack in the past 12 months.

676	People	Percentage of population affected by notifiable diseases	This indicator reflects percentage occurrence of notifiable diseases as reported by health departments to the National Notifiable Diseases Surveillance System (NNDSS). A notifiable disease is one for which regular, frequent, and timely information regarding individual cases is considered necessary for the prevention and control of the disease (CDC, 2005b). The "notifiable diseases" included are chlamydia, coccidioidomycosis, cryptosporidiosis, Dengue virus, *Escherichia coli*, ehrlichiosis, giardiasis, gonorrhea, *Haemophilus influenzae*, hepatitis A, hepatitis B, hepatitis C, legionellosis, Lyme disease, malaria, meningococcal disease, mumps, pertussis (whooping cough), rabies, salmonellosis, shigellosis, spotted fever rickettsiosis/Rocky Mountain spotted fever, *Streptococcus pneumoniae*, syphilis, tuberculosis, varicella (chicken pox), and West Nile/meningitis/encephalitis.
680	Natural Environment	Ecological connectivity (percentage of area classified as hub or corridor)	This indicator reflects the percentage of the metropolitan area identified as a "hub" or "corridor." Hubs are large areas of important natural ecosystems such as the Okefenokee National Wildlife Refuge in Georgia and the Osceola National Forest in Florida. Corridors (i.e., "connections") are links to support the functionality of the hubs (e.g., the Pinhook Swamp which connects the Okefenokee and Osceola hubs).
681	Natural Environment	Relative ecological condition of undeveloped land	This indicator characterizes the ecological condition of undeveloped land based on three indices derived from criteria representing diversity, self-sustainability, the rarity of certain types of land cover, species, and higher taxa (White and Maurice, 2004). In this context, "undeveloped land" refers to all land use not classified as urban, industrial, residential, or agricultural.
682	Natural Environment	Percentage change in bird population	This indicator reflects the number of species with "substantial" increases or decreases in the number of observations (not a change in the number of species) divided by the total number of bird species.
690	People	Emergency medical service response times	This indicator measures average annual response times (in minutes) for emergency medical service calls.
709	Economy	Percentage of owned housing units that are affordable	This indicator measures (1) the percentage of owned housing units where selected monthly ownership costs (rent, mortgages, real estate taxes, insurance, utilities, fuel, fees) as a percentage of household income (SMOCAPI) exceeds 35% or (2) the percentage of rented housing units where gross rent as a percentage of household income (GRAPI) exceeds 35%.

711	Economy	Overall unemployment rate	Employment is a measure of economic viability and self-sufficiency. Employment opportunities spread across different industries create a more stable employment base. A diversification of industries also offers opportunities to a diverse labor market. This indicator measures the percentage of sectors in a city's economy that employ < 40% of the city's population. Sectors that employ 1% or less of the city's population are not considered, as they provide very minimal employment opportunities.
717	Economy	Percentage access to health insurance of noninstitutionalized population	This indicator measures the percentage of noninstitutionalized residents with health insurance.
722	Economy	Percentage change in homeless population	This indicator measures the percentage change in the homeless population.
725	People	Number of physicians per capita	This indicator reflects the total number of M.D. and D.O. physicians per capita.
728	People	Adult care (homes per capita)	The number of adult day care homes and assisted living homes per capita of population over 65 years.
757	People	Average police response time	This indicator reflects the average response time for police to respond to emergency situations.
784	People	Number of sworn police officers per capita	This indicator is calculated by dividing the number of sworn police officers by the total population. We multiply the result by 1,000. According to the FBI, sworn officers meet the following criteria: "they work in an official capacity, they have full arrest powers, they wear a badge (ordinarily), they carry a firearm (ordinarily), and they are paid from governmental funds set aside specifically for payment of sworn law enforcement representatives." In counties with relatively few people, a small change in the number of officers may have a significant effect on rates from year to year.
798	People	Percentage of fire response times less than 6.5 minutes	This indicator reflects the percentage of fire response times less than 6.5 minutes (from city stations to city locations).
825	Land Use/Land Cover	Percent change in impervious cover	This indicator reflects the change in the percentage of the metropolitan area that is impervious surface (roads, buildings, sidewalks, parking lots, etc.).

898	Energy	Annual energy consumption per capita by main use category (commercial use)	The indicator measures the annual energy consumption (2010) per capita within the commercial use sector.
924	Energy	Energy intensity by use	This indicator measures energy intensity in manufacturing, transportation, agriculture, commercial and public services, and the residential sector.
949	Energy	Percentage energy consumed for electricity	The indicator measures electricity consumption per year in kWh as a percentage of total energy consumption.
950	Energy	Percentage of electricity generation from noncarbon sources	This indicator measures the percentage of total electricity generation from noncarbon energy sources in a city.
951	Energy	Percentage of total energy use from renewable sources	This indicator measures the percentage of total energy use from renewable sources.
967	Energy	Total energy source capacity per capita	This indicator measures the total capacity of all energy sources (MW) per capita.
970	Energy	Average capacity of a decentralized energy source	This indicator measures the average capacity of a decentralized energy source (m^3/acre). Decentralized energy sources are those that can be used as a supplementary source to the existing centralized energy system. They are typically located closer to the site of actual energy consumption than centralized sources.
971	Energy	Energy source capacity per unit area	This indicator measures the total capacity of energy sources per unit area served (MW/sq mi).
983	Energy	Average customer energy outage (hours) in recent major storm	This indicator measures the average customer energy outage hours divided by number of electricity customers for a storm event in June 2012.
985	Transportation	Transport system user satisfaction	This indicator reflects the overall user satisfaction with the transport system. It is defined as the average user satisfaction with bus service, rail service, and the accuracy of passenger information displays.
987	Transportation	Employment accessibility (mean travel time to work relative to national average)	This indicator is defined as the mean travel time to work in a city relative to the U.S. average.

988	Transportation	Walkability score	This indicator reflects the walkability score of the community (points out of 100).
991	Transportation	Percentage transport diversity	Highest public expenditure for a single mode of transportation as a percentage of the total expenditures for all transportation modes.
1003	Transportation	Mobility management (yearly congestion costs saved by operational treatments per capita)	This indicator reports on the yearly congestion costs saved by operational treatments (in billions of 2011 dollars). Operational treatments include freeway incident management, freeway ramp metering, arterial street signal coordination, arterial street access management, and high-occupancy vehicle lanes.
1010	Transportation	Community Livability Index	The Community Livability Index is the equally weighted average of the Community Service Indicator, the Crime Indicator, the Retail Opportunity Indicator, the Educational Indicator, the Environmental Quality Indicator, the Housing Affordability Indicator, and the Transit Livability Indicator. Details of the calculation are provided in Ripplinger et al. (2012; http://www.ugpti.org/pubs/pdf/DP262.pdf).
1157	People	Percentage of housing units with air conditioning	This indicator reflects the percentage of housing units with air conditioning.
1170	People	Percentage of population experiencing heat-related deaths	This indicator reflects the percentage of the population experiencing heat-related deaths.
1171	People	Percentage of population affected by food poisoning	This indicator reflects the percentage of population affected by food poisoning (i.e., *Salmonella* spp., unsafe drinking water).
1346	Water	Percentage of infiltration and inflow (I/I) in wastewater	Water that enters the wastewater system through infiltration and inflow (I/I) as a percentage of total wastewater from all wastewater treatment plants in the city. Infiltration is the seepage of groundwater into sewer pipes through cracks, holes, joint failures, or faulty connections. Inflow is surface water that enters the wastewater system from yard, roof, and footing drains; cross-connections with storm drains and downspouts; and through holes in manhole covers.
1347	Water	Wet weather flow bypass volume relative to the 5-year average	Volume of wastewater that bypassed treatment in an average year for all wastewater treatment plants divided by the 5-year average.
1369	Water	Annual coefficient of variation (CV) of unregulated streamflow	The coefficient of variation (CV) of unregulated streamflow is an indicator of annual streamflow variability. It is computed as the ratio of the standard deviation of unregulated annual streamflow (σQ_s) to

				the unregulated mean annual streamflow (QS)'. (Hurd et al., 1999).
1375	Economy	Percentage of population living below the poverty line		This indicator measures the percentage of the population living below the poverty line.
1376	People	Percentage of population that is disabled		This indicator reflects the percentage of the noninstitutionalized population that is disabled. Disabled individuals are those who have one or more of the following: hearing difficulty (deaf or having serious difficulty hearing), vision difficulty (blind or having serious difficulty seeing, even when wearing glasses), cognitive difficulty (having difficulty remembering, concentrating, or making decisions because of a physical, mental, or emotional problem), ambulatory difficulty (serious difficulty walking or climbing stairs), self-care difficulty (difficulty bathing or dressing), and independent living difficulty (difficulty doing errands because of a physical, mental, or emotional problem).
1387	People	Percentage of population vulnerable due to age		This indicator reflects percentage of population above 65 or under 5 years old.
1390	People	Percentage of population that is living alone		This indicator reflects the percentage of population that is 65 years or older and living alone.
1396	Transportation	Percent access to transportation stops		This indicator reflects the percentage of the population that is near a transit stop.
1399	Transportation	Percentage of roads and railroads within the city that are located within 10 feet of water		This indicator measures the percentage of roadway miles and rail line miles that are within 10 feet of a body of water.
1400	Transportation	Percentage of roads and railroads within the city in the 500-year floodplain		This indicator measures the percentage of roadway miles and rail line miles that are within the 500-year floodplain.
1401	Transportation	Percentage of roads and railroads within the city in the 100-year floodplain		This indicator measures the percentage of roadway miles and rail line miles that are within the 100-year floodplain.
1402	Transportation	Total annual hours of rail line closure due to heat and maintenance problems		This indicator measures (1) total annual hours that rail lines within the metropolitan transit system are closed due to heat kinks and (2) total annual hours that transit vehicles are unable to operate due to maintenance problems associated with extreme heat stress.

1403	Transportation	Percentage of city culverts that are sized to meet current stormwater capacity requirements	This indicator measures the percentage of current culverts that cross transportation facilities in the metropolitan region that are sized to meet current stormwater capacity requirements.
1404	Transportation	Percentage of city culverts that are sized to meet future stormwater capacity requirements	This indicator measures the percentage of current culverts that cross transportation facilities in the metropolitan region that are sized to meet projected stormwater capacity requirements for 2030.
1406	Transportation	Percentage decline in repeat maintenance events	This indicator measures the percentage decline in repeat maintenance events, thereby representing a stable transportation system. The most recent transportation bill states that roadways and bridges subject to repeat maintenance must be studied so as to avoid repeated use of emergency funds for infrastructure that keeps getting damaged.
1408	Transportation	Percentage of bridges that are structurally deficient	This indicator measures the percentage of bridges that are structurally deficient. Bridges are considered structurally deficient if significant load-carrying elements are found to be in poor or worse condition due to deterioration or damage, or the adequacy of the waterway opening provided by the bridge is determined to be extremely insufficient to the point of causing intolerable traffic interruptions.
1410	Transportation	Average Hours of passenger transit delay per capita due to heat related issues	This indicator measures the average hours of passenger transit delay per capita due to heat related issues.
1411	Transportation	Roadway connectivity (number of intersections per square mile)	This indicator measures the number of intersections per square mile.
1412	Transportation	Miles of pedestrian facilities per street mile	This indicator measures the miles of pedestrian facilities (sidewalks) per street mile.
1413	Transportation	Percentage of short walkable sidewalks in urban areas	This indicator measures the percentage of sidewalks within the urban area that are less than 330 feet.
1417	Transportation	Percentage funding spent on pedestrian/bicycle projects connected to community activity centers	Percentage of program funds spent on pedestrian or bicycle projects that include at least one connection to activity centers (e.g., schools; universities; downtown and employment districts; senior facilities; hospital/medical clinics; parks, recreation, and sporting; grocery stores; museums and tourist attractions).

1419	Transportation	Intermodal freight connectivity (ratio of intermodal connections used per year to individual modes)	This indicator measures the number of intermodal connections per year relative to distinct modes. Intermodal connections allow freight to use a combination of modes and give shippers additional transportation alternatives that unconnected, parallel systems do not offer.
1420	Transportation	Intermodal passenger connectivity (percentage of terminals with at least one intermodal connection for the most common mode)	This indicator measures the percentage of active passenger terminals for the most common mode (e.g., rail, air) with at least one intermodal passenger connection. Intermodal connections allow passengers to use a combination of modes and give travelers additional transportation alternatives that unconnected, parallel systems do not offer.
1422	Transportation	Average distance of all nonwork trips	This indicator measures the average distance from a given home to the nearest grocery store, high school, and health care facility (i.e., nonwork trips).
1424	Transportation	Roundabouts	N/A
1426	Transportation	City congestion rank	This indicator measures the congestion rank of the metropolitan area relative to all U.S. metropolitan areas.
1428	Water	Total number of Safe Drinking Water Act (SDWA) violations	This indicator measures the total number of SDWA violations over the last 5 years.
1429	Transportation	Telework rank	This indicator measures the telework rank of the metropolitan area relative to all other extra-large metropolitan areas in the U.S. The rank is based on the percentage of jobs within the metropolitan region that could be accomplished by telecommuting if employer policies were to permit it.
1433	Telecommunications	Percentage of system capacity needed to carry baseline level of traffic	N/A
1434	Telecommunications	Baseline percentage of water supply for telecommunication systems that comes from outside the metropolitan area	N/A

1435	Telecommunications	Baseline percentage of energy supply for telecommunication systems that comes from outside the metropolitan area	N/A
1436	Land Use/Land Cover	Percentage of city area in 100-year floodplain	This indicator reflects the percentage of the metropolitan area that lies within the 100-year floodplain.
1437	Land Use/Land Cover	Percentage of city area in 500-year floodplain	This indicator reflects the percentage of the metropolitan area that lies within the 500-year floodplain.
1438	Land Use/Land Cover	Percentage of city population in 100-year floodplain	This indicator reflects the percentage of the city population living within the 100-year floodplain.
1439	Land Use/Land Cover	Percentage of city population in 500-year floodplain	This indicator reflects the percentage of the city population living within the 500-year floodplain.
1440	Land Use/Land Cover	Palmer Drought Severity Index	Calculate potential evapotranspiration (PET) for selected time periods using temperature data and the Thornthwaite equation. Find the precipitation deficit (precipitation minus PET) for the selected time period, where more negative values indicate greatest precipitation deficit. Using a moving window sum, find the 1-, 3-, 6-, or 12-month period that had the greatest total precipitation deficit.
1441	Telecommunications	Percentage of community with access to FEMA emergency radio broadcasts	Percentage of community with access to FEMA emergency radio broadcasts.
1442	Water	Ratio of water consumption to water availability	This indicator measures the fraction of available water that is currently consumed. It is calculated by dividing the total available water from surface water and groundwater sources by total water consumption.
1443	People	Deaths from extreme weather events	This indicator measures the number of deaths in the last 5 years due to extreme events (cold, flood, heat, lightning, tornado, tropical cyclone, wind, and winter storms).

APPENDIX I. QUALITATIVE INDICATORS: TEMPLATE

A complete set of the qualitative indicators by sector developed for the tool.

I.1. Economy

The questions below have been developed for the economy sector. Each question is flagged with one or more of the following gradual change climate stressor and/or extreme event climate stressor (from the urban resilience framework developed for this project):

Stressors

Gradual Changes
- Wind speed
- Temperature
- Precipitation
- Sea level rise

Extreme Events
- Magnitude/duration of heat waves
- Drought intensity/duration
- Flood magnitude/frequency
- Hurricane intensity/frequency
- Storm surge/flooding

In addition, each question has up to four possible **answers**. Each answer has been assigned a **resilience score** on a scale of 1 (lowest resilience) to 4 (highest resilience).

For each question, please:

1. Discuss the **relevance** of the question to the economy sector. (If unsure, please select the *not sure—remind me later* option.) Questions may be selected as *yes (relevant)* on the basis of the stressors previously selected as being most relevant to Washington, DC or based on any other criteria.
2. For questions marked as *yes (relevant)*, discuss an **importance weight**, where 1 = not very important and 4 = very important.
3. For questions marked as *yes (relevant)*, identify the best **answer** to the question from the options provided.

#1: Is the economy of the urban area largely independent, or is it largely dependent on economic activity in other urban areas?

Relevance	*Importance Weights*
Yes (relevant)	1 (not very important)
No (not relevant)	2
Not sure—remind me later	3
	4 (very important)
Answer	*Resilience Score*
Largely dependent	1 (lowest resilience)
Somewhat dependent	2
Somewhat independent	3
Largely independent	4 (highest resilience)

#2: Does the urban area have mechanisms to help businesses quickly return to normal operations?

Relevance	*Importance Weights*
Yes (relevant)	1 (not very important)
No (not relevant)	2
Not sure—remind me later	3
	4 (very important)
Answer	*Resilience Score*
No	1 (lowest resilience)
Yes	3 (highest resilience)

#3: If jobs are lost in one sector of the urban area, does the capacity exist to expand the economy and job opportunities in another sector?

Relevance	*Importance Weights*
Yes (relevant)	1 (not very important)
No (not relevant)	2
Not sure—remind me later	3
	4 (very important)
Answer	*Resilience Score*
No	1 (lowest resilience)
Yes	3 (highest resilience)

#4: Has the vulnerability of critical infrastructure been assessed? Are there plans to relocate or protect vulnerable infrastructure in ways that promote resilience and protect other infrastructure and properties?

Relevance	*Importance Weights*
Yes (relevant)	1 (not very important)
No (not relevant)	2
Not sure—remind me later	3
	4 (very important)
Answer	*Resilience Score*
Vulnerability has not been assessed and there are no plans to protect infrastructure in ways that promote resilience.	1 (lowest resilience)
Vulnerability may or may not have been assessed, but infrastructure is insufficiently protected.	2
Yes, vulnerability has been assessed and infrastructure is somewhat protected in ways that promote resilience.	3
Yes, vulnerability has been assessed and infrastructure is protected in ways that promote resilience.	4 (highest resilience)

#5: Has the urban area's resilience to major changes in energy policy/prices been assessed?

Relevance	*Importance Weights*
Yes (relevant)	1 (not very important)
No (not relevant)	2
Not sure—remind me later	3
	4 (very important)

Answer	*Resilience Score*
No	1 (lowest resilience)
Yes	3 (highest resilience)

#6: Is funding available for adaptive development projects that could also serve as recreation areas (e.g., retention areas along waterways that could also serve as parks)? Are such multipurpose projects required or are there incentives for these projects?

Relevance	*Importance Weights*
Yes (relevant)	1 (not very important)
No (not relevant)	2
Not sure—remind me later	3
	4 (very important)

Answer	*Resilience Score*
No funding is available for these adaptive development projects and requirements or incentives do not exist for these projects.	1 (lowest resilience)
Funding is available for these adaptive development projects and requirements or incentives exist for these projects.	3 (highest resilience)

#7: Is a significant portion of the population of the urban area either seasonal residents or transient populations that may have a lesser degree of understanding of changes occurring within that area?

Relevance	*Importance Weight*
Yes	3

Answer	*Resilience Score*
Yes	1 (lowest resilience)
No	3 (highest resilience)

#8: How many people are in place to respond to emergencies, and what is the level of communication connectivity of emergency response teams and offices?

Relevance	*Importance Weights*
Yes (relevant)	1 (not very important)
No (not relevant)	2
Not sure—remind me later	3
	4 (very important)

Answer	*Resilience Score*
Many fewer people than necessary are in place for emergency response relative to urban area population, and communication connectivity teams and offices is poor.	1 (lowest resilience)
Too few people than necessary are in place for emergency response relative to urban area population, and communication connectivity teams and offices is fair.	2
Enough people are in place for emergency response relative to urban area population, and communication connectivity teams and offices is good.	3
A large number of people are in place for emergency response relative to urban area population, and communication connectivity teams and offices is excellent.	4 (highest resilience)

#9: Is comprehensive adaptation planning possible with the urban area's current resources? If so, is adaptation planning already occurring?

Relevance	*Importance Weights*
Yes (relevant)	1 (not very important)
No (not relevant)	2
Not sure—remind me later	3
	4 (very important)

Answer	*Resilience Score*
Resources do not allow for comprehensive adaptation planning.	1 (lowest resilience)
Resources would allow for adaptation planning, but no adaptation planning is occurring.	2
Some adaptation planning is occurring.	3
A great deal of adaptation planning is occurring.	4 (highest resilience)

#10: Is planning for climate change adaptation in the urban area incorporated into one office within the local government or is planning spread out across several offices within the government?

Relevance	*Importance Weight*

Answer	*Resilience Score*
Adaptation planning responsibilities are not incorporated into any offices within the local government.	1 (lowest resilience)
Adaptation planning responsibilities are spread out over multiple offices within the local government.	2
Adaptation planning is shared between two or three offices within the local government.	3
Adaptation planning is incorporated into one office within the local government.	4 (highest resilience)

#11: How flexible are planning processes for short-term and long-term responses? For example, is there flexibility in changing planning priorities if necessary?

Relevance	Importance Weights
Yes (relevant)	1 (not very important)
No (not relevant)	2
Not sure—remind me later	3
	4 (very important)
Answer	Resilience Score
Planning processes are fairly inflexible.	1 (lowest resilience)
Planning processes are somewhat flexible.	2
Planning processes are moderately flexible.	3
Planning processes are very flexible.	4 (highest resilience)

#12: Does adaptation planning for the urban area include retrospective analyses of past events (including analyses of past climate events in other cities if helpful) to help determine whether decisions on adaptation measures would be effective?

Relevance	Importance Weight
Yes (relevant)	1 (not very important
No (not relevant)	2
Not sure—remind me later	3
	4 (very important)
Answer	Resilience Score
Adaptation planning does not involve analyses of past climate-related events OR adaptation planning is not occurring.	1 (lowest resilience)
Adaptation planning occasionally involves analyses of past climate-related events.	2
Adaptation planning sometimes involves analyses of past climate-related events.	3
Adaptation planning frequently involves analyses of past climate-related events.	4 (highest resilience)

#13: Does adaptation planning for the urban area consider the costs and benefits of possible decisions, and does it encourage both pre-event and postevent evaluations of the effectiveness of adaptation measures?

Relevance	*Importance Weights*
Yes (relevant)	1 (not very important)
No (not relevant)	2
Not sure—remind me later	3
	4 (very important)

Answer	*Resilience Score*
Adaptation planning does not consider costs and benefits and does not encourage pre-event or postevent effectiveness evaluations.	1 (lowest resilience)
Adaptation planning does consider costs and benefits but does not encourage pre-event or postevent effectiveness evaluations.	2
Adaptation planning does consider costs and benefits and encourages pre-event or postevent effectiveness evaluations.	3
Adaptation planning does consider costs and benefits and requires pre-event or postevent effectiveness evaluations.	4 (highest resilience)

#14: Do adaptation plans account for tradeoffs between the less resilient but lower cost strategy of increasing protection from climatic changes and the more resilient but higher cost strategy of moving residents from the most vulnerable portions of the urban area? (One example of such a tradeoff is: in coastal cities, some areas can be protected by a seawall, or households and institutions in vulnerable areas can be moved inland. Do current adaptation plans account for the resilience-cost tradeoffs in this decision?)

Relevance	*Importance Weight*
Yes (relevant)	1 (not very important)
No (not relevant)	2
Not sure—remind me later	3
	4 (very important) |//
Answer	*Resilience Score*
Adaptation plans do not explicitly consider resilience-cost tradeoffs or no adaptation plans exist.	1 (lowest resilience)
Adaptation plans consider one or two resilience-cost tradeoffs.	2
Adaptation plans consider some resilience-cost tradeoffs.	3
Adaptation plans consider many resilience-cost tradeoffs.	4 (highest resilience)

#165: What financial capacity or credit risk is indicated by the city's bond rating(s)?

Relevance	*Importance Weight*
Yes (relevant)	1 (not very important)
No (not relevant)	2
Not sure—remind me later	3
	4 (very important)

Answer	*Resilience Score*
The bond rating(s) indicate(s) high vulnerability or very high credit risk/default.	1 (lowest resilience)
The bond rating(s) indicate(s) some vulnerability or substantial to high credit risk.	2
The bond rating(s) indicate(s) adequate financial capacity or some credit risk.	3
The bond rating(s) indicate(s) strong financial capacity/minimal to low credit risk.	4 (highest resilience)

I.2. Energy

The questions below have been developed for the energy sector. Each question is flagged with one or more of the following gradual change climate stressor and/or extreme event climate stressor (from the urban resilience framework developed for this project):

Stressors

Gradual Changes

- Wind speed
- Temperature
- Precipitation
- Sea level rise

Extreme Events

- Magnitude/duration of heat waves
- Drought intensity/duration
- Flood magnitude/frequency
- Hurricane intensity/frequency
- Storm surge/flooding

In addition, each question has up to four possible **answers**. Each answer has been assigned a **resilience score** on a scale of 1 (lowest resilience) to 4 (highest resilience).

For each question, please:

1. Discuss the **relevance** of the question to the energy sector. (If unsure, please select the *not sure—remind me later* option.) Questions may be selected as *yes (relevant)* on the basis of the stressors previously selected as being most relevant to Washington, DC or based on any other criteria.
2. For questions marked as *yes (relevant)*, discuss an **importance weight**, where 1 = not very important and 4 = very important.
3. For questions marked as *yes (relevant)*, identify the best **answer** to the question from the options provided.

#15: Do you have a diverse energy portfolio?

Relevance	Importance Weights
Yes (relevant)	1 (not very important)
No (not relevant)	2
Not sure—remind me later	3
	4 (very important)

Answer	Resilience Score
No	1 (lowest resilience)
Yes	2
	3
	4 (highest resilience)

#16: Are there redundant systems in place for coping with extreme events?

Relevance	Importance Weights
Yes (relevant)	1 (not very important)
No (not relevant)	2
Not sure—remind me later	3
	4 (very important)

Answer	Resilience Score
No, redundant energy systems are not in place.	1 (lowest resilience)
Yes, but these redundant energy systems have only a small amount of the capacity necessary.	2
Yes, and these redundant energy systems have some of the capacity necessary.	3
Yes, and these redundant energy systems have all the capacity necessary.	4 (highest resilience)

#17: To what extent do energy supplies come from outside the metropolitan area?

Relevance	Importance Weights
Yes (relevant)	1 (not very important)
No (not relevant)	2
Not sure—remind me later	3
	4 (very important)

Answer	Resilience Score
They come exclusively from outside the area.	1 (lowest resilience)
To a great extent	2
To a moderate extent	3
Only to a small extent	4 (highest resilience)

#18: Is the availability of energy goods and services at risk if other city goods and services (e.g., water, transportation, telecommunications) are affected by extreme climatic events or gradual climatic changes?

Relevance	Importance Weights
Yes (relevant)	1 (not very important)
No (not relevant)	2
Not sure—remind me later	3
	4 (very important)

Answer	Resilience Score
Availability of energy resources is at significant risk if other city services are affected by climatic events or changes.	1 (lowest resilience)
Availability of energy resources is at moderate risk if other city services are affected by climatic events or changes.	2
Availability of energy resources is at some risk if other city services are affected by climatic events or changes.	3
Availability of energy resources is at minimal risk if other city services are affected by climatic events or changes.	4 (highest resilience)

#19: How many minutes per year or hours per year do you have power outages?

Relevance	Importance Weight
Yes (relevant)	1 (not very important)
No (not relevant)	2
Not sure—remind me later	3
	4 (very important)

Answer	Resilience Score
More than 1 day per year for all outage events	1 (lowest resilience)
More than 1 hour to 1 day per year for all outage events	2
More than 30 minutes to 1 hour per year for all outage events	3
Less than 30 minutes per year for all outage events	4 (highest resilience)

#20: What is the response time to restore electrical power after an outage?

Relevance	Importance Weights
Yes (relevant)	1 (not very important)
No (not relevant)	2
Not sure—remind me later	3
	4 (very important)

Answer	Resilience Score
More than 1 day after a major event	1 (lowest resilience)
More than 3 hours to 1 day after a major event	2
More than 1 hour to 4 hours after a major event	3
Less than 1 hour after a major event	4 (highest resilience)

#21: Does capacity exist to handle a higher peak demand or peaks at different times?

Relevance	*Importance Weights*
Yes (relevant)	1 (not very important)
No (not relevant)	2
Not sure—remind me later	3
	4 (very important)

Answer	*Resilience Score*
Electricity generation capacity cannot handle higher peak demands or peaks at different times than currently experienced.	1 (lowest resilience)
Electricity generation capacity can handle higher peak demands or peaks at different times than currently experienced.	3 (highest resilience)

#22: To what extent have efforts been made to reduce energy demand?

Relevance	*Importance Weights*
Yes (relevant)	1 (not very important)
No (not relevant)	2
Not sure—remind me later	3
	4 (very important)

Answer	*Resilience Score*
Few to no efforts have been made to reduce energy demand.	1 (lowest resilience)
Fair efforts have been made to reduce energy demand.	2
Moderate efforts have been made to reduce energy demand.	3
Significant efforts have been made to reduce energy demand.	4 (highest resilience)

#23: What are the opportunities for distributed generation sources (i.e., different capacity for energy generation from different sources including renewable)?

Relevance	*Importance Weights*
Yes (relevant)	1 (not very important)
No (not relevant)	2
Not sure—remind me later	3
	4 (very important)

Answer	*Resilience Score*
Political and technical capacity do not allow for generation from multiple sources.	1 (lowest resilience)
Political and technical capacity could allow for generation from multiple sources, but such diversified generation is not currently occurring.	2
Political and technical capacity currently provide for generation from multiple sources, not including renewables.	3
Political and technical capacity currently provide for generation from multiple sources, including renewables.	4 (highest resilience)

#24: Are there smart grid opportunities to manage demand?

Relevance	*Importance Weights*
Yes (relevant)	1 (not very important)
No (not relevant)	2
Not sure—remind me later	3
	4 (very important)

Answer	*Resilience Score*
No	1 (lowest resilience)
Yes	3 (highest resilience)

#147: Do municipal managers draw on past data/experiences of extreme weather events to assess the effects of these events on oil and gas availability and pricing? (DOE, 2013)

Relevance	Importance Weight
Yes (relevant)	1 (not very important)
No (not relevant)	2
Not sure—remind me later	3
	4 (very important)

Answer	Resilience Score
No	1 (lowest resilience)
Yes	3 (highest resilience)

#148: Has the city consulted with local power companies to develop plans for potential increases in electricity demand for summer cooling? (DOE, 2013)

Relevance	Importance Weight
Yes (relevant)	1 (not very important)
No (not relevant)	2
Not sure—remind me later	3
	4 (very important)

Answer	Resilience Score
The city has not consulted with local power companies and is not developing plans for potential increase in electricity for cooling.	1 (lowest resilience)
The city has consulted with local power companies regarding potential increase in electricity for cooling, but is not yet developing related plans. OR the city has developed such plans, but did not consult with local power companies.	2
The city has consulted with local power companies and is developing plans for potential increase in electricity for cooling.	3
The city has consulted with local power companies and developed plans for potential increase in electricity for cooling.	4 (highest resilience)

#149: Has the city coordinated with local water suppliers and power generation facilities to discuss potential climate-induced water shortages and their impacts on cooling the power generation facilities?(DOE, 2013)

Relevance	*Importance Weight*
Yes (relevant)	1 (not very important)
No (not relevant)	2
Not sure—remind me later	3
	4 (very important)

Answer	*Resilience Score*
No	1 (lowest resilience)
Yes	3 (highest resilience)

#150: Do municipal managers in coastal areas consider the impacts of sea level rise on power generation facilities?

Relevance	*Importance Weight*
Yes (relevant)	1 (not very important)
No (not relevant)	2
Not sure—remind me later	3
	4 (very important)

Answer	*Resilience Score*
No	1 (lowest resilience)
Yes, but these considerations are not incorporated into planning for these facilities.	2
Yes, and these considerations are being incorporated into planning for these facilities.	3
Yes, and these considerations are incorporated into planning for these facilities.	4 (highest resilience)

I.3. Land Use/Land Cover

The questions below have been developed for the land use/land cover sector. Each question is flagged with one or more of the following gradual change climate stressor and/or extreme event climate stressor (from the urban resilience framework developed for this project):

Stressors

Gradual Changes
- Wind speed
- Temperature
- Precipitation
- Sea level rise

Extreme Events
- Magnitude/duration of heat waves
- Drought intensity/duration
- Flood magnitude/frequency
- Hurricane intensity/frequency
- Storm surge/flooding

In addition, each question has up to four possible **answers**. Each answer has been assigned a **resilience score** on a scale of 1 (lowest resilience) to 4 (highest resilience).

For each question, please:

1. Discuss the **relevance** of the question to the land use/land cover sector. (If unsure, please select the *not sure—remind me later* option.) Questions may be selected as *yes (relevant)* on the basis of the stressors previously selected as being most relevant to Washington, DC or based on any other criteria.
2. For questions marked as *yes (relevant)*, discuss an **importance weight**, where 1 = not very important and 4 = very important.
3. For questions marked as *yes (relevant)*, identify the best **answer** to the question from the options provided.

#25: Can resilience planning/adaptation be incorporated into existing programs that communities engage in regularly (e.g., zoning, hazard mitigation plans)?

Relevance	*Importance Weights*
Yes (relevant)	1 (not very important)
No (not relevant)	2
Not sure—remind me later	3
	4 (very important)

Answer	*Resilience Score*
Resilience planning/adaptation would be difficult to incorporate in regular planning programs.	1 (lowest resilience)
Resilience planning/adaptation could be incorporated in regular planning programs, but this may be difficult.	2
Resilience planning/adaptation could be incorporated in regular planning programs with some effort.	3
Resilience planning/adaptation is incorporated in regular planning programs.	4 (highest resilience)

#26: Has the city made efforts to use urban forms to mitigate climate change impacts and to maximize benefits (e.g., urban tree canopy cover)?

Relevance	*Importance Weights*
Yes (relevant)	1 (not very important)
No (not relevant)	2
Not sure—remind me later	3
	4 (very important)

Answer	*Resilience Score*
The city is not considering and has not developed efforts to use urban form to mitigate climate change impacts and maximize the benefits of urban forms.	1 (lowest resilience)
The city is considering development of efforts to use urban form to mitigate climate change impacts and maximize the benefits of urban forms.	2
The city is developing efforts to use urban form to mitigate climate change impacts and maximize the benefits of urban forms.	3
The city has developed and implemented efforts to use urban form to mitigate climate change impacts and maximize the benefits of urban forms.	4 (highest resilience)

#27: Are urban forms used that address (lessen) urban heat island effects (e.g., through increasing evapotranspiration or increasing urban ventilation)?

Relevance	*Importance Weights*
Yes (relevant)	1 (not very important)
No (not relevant)	2
Not sure—remind me later	3
	4 (very important)

Answer	*Resilience Score*
These forms are not used in new development and retrofits/renovations of old development.	1 (lowest resilience)
These forms are infrequently used in new development and retrofits/renovations of old development.	2
These forms are sometimes used in new development and retrofits/renovations of old development.	3
These forms are often used in new development and retrofits/renovations of old development.	4 (highest resilience)

#28: Does zoning encourages green roofs or other practices that reduce urban heat?

Relevance	*Importance Weights*
Yes (relevant)	1 (not very important)
No (not relevant)	2
Not sure—remind me later	3
	4 (very important)

Answer	*Resilience Score*
Zoning does not allow green roofs and other practices that reduce the urban heat island effect.	1 (lowest resilience)
Zoning discourages green roofs and other practices that reduce the urban heat island effect.	2
Zoning allows green roofs and other practices that reduce the urban heat island effect.	3
Zoning encourages green roofs and other practices that reduce the urban heat island effect.	4 (highest resilience)

#29: Are there mechanisms to support tree shading programs in urban areas (to reduce urban heat and improve air quality)? Are there innovative ways to fund such programs?

Relevance	Importance Weights
Yes (relevant)	1 (not very important)
No (not relevant)	2
Not sure—remind me later	3
	4 (very important)

Answer	Resilience Score
No, such mechanisms do not exist.	1 (lowest resilience)
Yes, there are such mechanisms; additional funding is needed, likely through new or innovative sources.	2
Yes, there are such mechanisms; additional funding is needed but could be provided through existing sources.	3
Yes, there are such mechanisms, and they are well funded.	4 (highest resilience)

#30: Have land use/land cover types, such as soil and vegetation types and areas of tree canopy cover, been inventoried, and are these inventories used in planning?

Relevance	Importance Weights
Yes (relevant)	1 (not very important)
No (not relevant)	2
Not sure—remind me later	3
	4 (very important)

Answer	Resilience Score
Land use/land cover types are not inventoried and are not planned to be inventoried.	1 (lowest resilience)
Plans exist to inventory land use/land cover types OR inventories exist but existing inventories are not used in planning.	2
Land use/land cover types are being inventoried and these inventories are used or will be used in planning.	3
Land use/land cover types have been inventoried and these inventories are used in planning.	4 (highest resilience)

#31: What percentage of open/green space is required for new development (to encourage increases in such space)?

Relevance	Importance Weights
Yes (relevant)	1 (not very important)
No (not relevant)	2
Not sure—remind me later	3
	4 (very important)

Answer	Resilience Score
No open/green space is required for new development.	1 (lowest resilience)
A small percentage of open/green space is required for new development.	2
A moderate percentage of open/green space is required for new development.	3
A high percentage of open/green space is required for new development.	4 (highest resilience)

#32: Are there mechanisms for the local government to purchase land that is unfavorable for redevelopment due to the results of extreme events (e.g., flooding from a hurricane)? If so, what are those mechanisms?

Relevance	Importance Weights
Yes (relevant)	1 (not very important)
No (not relevant)	2
Not sure—remind me later	3
	4 (very important)

Answer	Resilience Score
No, such mechanisms do not exist.	1 (lowest resilience)
Yes, there are such mechanisms, but they are only preliminary and are slightly helpful.	2
Yes, there are such mechanisms and they are somewhat helpful.	3
Yes, there are such mechanisms and they are helpful.	4 (highest resilience)

#33: Are there policies or zoning practices in place that allow transfer of ownership of undevelopable land subject to flooding or excessive erosion to the city (or allow nonpermanent structures only)? Are these policies or zoning practices enforced?

Relevance	*Importance Weights*
Yes (relevant)	1 (not very important)
No (not relevant)	2
Not sure—remind me later	3
	4 (very important)

Answer	*Resilience Score*
Policies do not allow ownership transfer.	1 (lowest resilience)
Policies allow ownership transfer, but these policies are enforced only rarely.	2
Policies allow ownership transfer, but these policies are only enforced some of the time.	3
Policies allow ownership transfers, and these policies are enforced.	4 (highest resilience)

#34: Where developed land is located in areas vulnerable to extreme events, are resilient retrofits being implemented or planned?

Relevance	*Importance Weights*
Yes (relevant)	1 (not very important)
No (not relevant)	2
Not sure—remind me later	3
	4 (very important)

Answer	*Resilience Score*
No	1 (lowest resilience)
Yes	3 (highest resilience)

#35: Are there codes to prevent development in flood-prone areas?

Relevance	*Importance Weights*
Yes (relevant)	1 (not very important)
No (not relevant)	2
Not sure—remind me later	3
	4 (very important)

Answer	*Resilience Score*
No	1 (lowest resilience)
Yes	3 (highest resilience)

#36: Are there regulations in place regarding whether communities that are affected by floods will be rebuilt in the same location?

Relevance	*Importance Weights*
Yes (relevant)	1 (not very important)
No (not relevant)	2
Not sure—remind me later	3
	4 (very important)

Answer	*Resilience Score*
No, regulations do not exist regarding the location of rebuilding efforts for communities affected by floods.	1 (lowest resilience)
Regulations regarding location of rebuilding efforts for communities affected by floods.	2
Yes, regulations exist and encourage communities strongly affected by floods to rebuild using more flood-resistant structures and methods.	3
Yes, regulations exist and encourage communities strongly affected by floods to be rebuilt in locations less prone to flooding.	4 (highest resilience)

#37: Have the regulations regarding rebuilding of communities affected by floods been enforced to date?

Relevance	Importance Weights
Yes (relevant)	1 (not very important)
No (not relevant)	2
Not sure—remind me later	3
	4 (very important)

Answer	Resilience Score
No	1 (lowest resilience)
Yes	3 (highest resilience)

#38: Do incentives exist to integrate green stormwater infrastructure into infrastructure planning to mitigate flooding?

Relevance	Importance Weights
Yes (relevant)	1 (not very important)
No (not relevant)	2
Not sure—remind me later	3
	4 (very important)

Answer	Resilience Score
No	1 (lowest resilience)
Yes	3 (highest resilience)

#39: Are there incentives to reduce the amount of impervious surface, to prevent development in floodplains, to use urban forestry to reduce impacts, to use green infrastructure for stormwater management, etc.?

Relevance	*Importance Weights*
Yes (relevant)	1 (not very important)
No (not relevant)	2
Not sure—remind me later	3
	4 (very important)

Answer	*Resilience Score*
No, such incentives do not exist.	1 (lowest resilience)
Yes, incentives exist to promote green infrastructure-oriented solutions to stormwater management.	3 (highest resilience)

#40: To what extent was green infrastructure selected to provide the maximum ecological benefits?

Relevance	*Importance Weights*
Yes (relevant)	1 (not very important)
No (not relevant)	2
Not sure—remind me later	3
	4 (very important)

Answer	*Resilience Score*
Green infrastructure does not exist or green infrastructure does not provide ecological benefits.	1 (lowest resilience)
Green infrastructure was selected with minimal attention to the ecological benefits provided.	2
Green infrastructure was selected to provide some ecological benefits.	3
Green infrastructure was selected to provide the maximum ecological benefits.	4 (highest resilience)

#41: Has green infrastructure maintenance been built into the budget?

Relevance	Importance Weights
Yes (relevant)	1 (not very important)
No (not relevant)	2
Not sure—remind me later	3
	4 (very important)

Answer	Resilience Score
No	1 (lowest resilience)
Yes	3 (highest resilience)

#142: Are coastal hazard maps with 1-meter altitude contours available, and are these maps used in planning?

Relevance	Importance Weight
Yes (relevant)	1 (not very important)
No (not relevant)	2
Not sure—remind me later	3
	4 (very important)

Answer	Resilience Score
Such maps have not been developed and are not planned to be developed.	1 (lowest resilience)
Plans exist to develop such maps OR such maps exist but are not used in planning.	2
Such maps are being developed and these maps are used or will be used in planning.	3
Such maps exist and these maps are used in planning.	4 (highest resilience)

#151: Have institutional land practices (i.e., zoning, land use planning) potentially been hindered by other government agencies seeking to shift financial resources when it comes to climate change planning?

Relevance	Importance Weight
Yes (relevant)	1 (not very important)
No (not relevant)	2
Not sure—remind me later	3
	4 (very important)

Answer	Resilience Score
Yes	1 (lowest resilience)
No	3 (highest resilience)

#152: Does knowledge of historical land use/land cover changes contribute to planners' understanding of climate stresses?

Relevance	Importance Weight
Yes (relevant)	1 (not very important)
No (not relevant)	2
Not sure—remind me later	3
	4 (very important)

Answer	Resilience Score
No	1 (lowest resilience)
Yes	3 (highest resilience)

#153: Have specific historical land use/land cover changes been recognized as increasing or decreasing vulnerability to climate stresses?

Relevance	*Importance Weight*
Yes (relevant)	1 (not very important)
No (not relevant)	2
Not sure—remind me later	3
	4 (very important)

Answer	*Resilience Score*
No	1 (lowest resilience)
Yes	3 (highest resilience)

#154: Does the city consider the knowledge of local academic research and other stakeholders (e.g., farmers, forest managers, land use managers) in land use planning related to climate resilience?

Relevance	*Importance Weight*
Yes (relevant)	1 (not very important)
No (not relevant)	2
Not sure—remind me later	3
	4 (very important)

Answer	*Resilience Score*
No	1 (lowest resilience)
Yes	3 (highest resilience)

#167: In general, what is the monetary value of infrastructure located within the 500-year floodplain in the city?

Relevance	Importance Weight
Yes (relevant)	1 (not very important)
No (not relevant)	2
Not sure—remind me later	3
	4 (very important)

Answer	Resilience Score
The monetary value of infrastructure in the 500-year floodplain is high.	1 (lowest resilience)
The monetary value of infrastructure in the 500-year floodplain is moderate.	2
The monetary value of infrastructure in the 500-year floodplain is low.	3
The monetary value of infrastructure in the 500-year floodplain is very low.	4 (highest resilience)

I.4. Natural Environment

The questions below have been developed for the natural environment sector. Each question is flagged with one or more of the following gradual change climate stressor and/or extreme event climate stressor (from the urban resilience framework developed for this project):

Stressors

Gradual Changes	*Extreme Events*
Wind speed	Magnitude/duration of heat waves
Temperature	Drought intensity/duration
Precipitation	Flood magnitude/frequency
Sea level rise	Hurricane intensity/frequency
	Storm surge/flooding

In addition, each question has up to four possible **answers**. Each answer has been assigned a **resilience score** on a scale of 1 (lowest resilience) to 4 (highest resilience).

For each question, please:

1. Discuss the **relevance** of the question to the natural environment sector. (If unsure, please select the *not sure—remind me later* option.) Questions may be selected as *yes (relevant)* on the basis of the stressors previously selected as being most relevant to Washington, DC or based on any other criteria.
2. For questions marked as *yes (relevant)*, discuss an **importance weight**, where 1 = not very important and 4 = very important.
3. For questions marked as *yes (relevant)*, identify the best **answer** to the question from the options provided.

#42: Is the availability of environmental/ecosystem goods and services at risk if other city goods and services (e.g., power, water, telecommunications) are affected by extreme climatic events or gradual climatic changes?

Relevance	*Importance Weights*
Yes (relevant)	1 (not very important)
No (not relevant)	2
Not sure—remind me later	3
	4 (very important)
Answer	*Resilience Score*
Availability of environmental/ecosystem resources is at significant risk if other city services are affected by climatic events or changes.	1 (lowest resilience)
Availability of environmental/ecosystem resources is at moderate risk if other city services are affected by climatic events or changes.	2
Availability of environmental/ecosystem resources is at some risk if other city services are affected by climatic events or changes.	3
Availability of environmental/ecosystem resources is at minimal risk if other city services are affected by climatic events or changes.	4 (highest resilience)

#43: What regulatory and planning tools related to air quality, water quality, and land use are already available locally? For example, does the urban area have invasive plant ordinances or tree planting requirements?

Relevance	*Importance Weights*
Yes (relevant)	1 (not very important)
No (not relevant)	2
Not sure—remind me later	3
	4 (very important)

Answer	*Resilience Score*
The urban area does not have regulatory and planning tools for air and water quality and land use.	1 (lowest resilience)
The urban area has few regulatory and planning tools for air and water quality and land use.	2
The urban area has several regulatory and planning tools for air and water quality and land use.	3
The urban area has many regulatory and planning tools for air and water quality and land use.	4 (highest resilience)

#44: Do plans exist for increasing open and green space?

Relevance	*Importance Weights*
Yes (relevant)	1 (not very important)
No (not relevant)	2
Not sure—remind me later	3
	4 (very important)

Answer	*Resilience Score*
No	1 (lowest resilience)
Yes	3 (highest resilience)

#45: Has the continuity of open or green spaces been assessed and addressed in planning efforts?

Relevance	*Importance Weights*
Yes (relevant)	1 (not very important)
No (not relevant)	2
Not sure—remind me later	3
	4 (very important)

Answer	*Resilience Score*
Continuity of open or green spaces has not been assessed and is not planned to be assessed.	1 (lowest resilience)
Plans exist to assess the continuity of open or green spaces OR an assessment has been completed but is not addressed in planning efforts.	2
Continuity of open or green spaces is being assessed and is or will be addressed in planning efforts.	3
Continuity of open or green spaces has been assessed and is addressed in planning efforts.	4 (highest resilience)

#46: Do native plant or animal species lists exist for the urban area, and are these species (rather than nonnative species) used in green infrastructure?

Relevance	*Importance Weights*
Yes (relevant)	1 (not very important)
No (not relevant)	2
Not sure—remind me later	3
	4 (very important)

Answer	*Resilience Score*
Native species lists do not exist and are not being developed.	1 (lowest resilience)
Native species lists exist, but green infrastructure uses mostly nonnative species OR native species lists are under development.	2
Native species lists exist and green infrastructure uses mostly these species.	3
Native species lists exist and green infrastructure uses only these species.	4 (highest resilience)

#47: Does the urban area coordinate with other nearby entities on water quality?

Relevance	*Importance Weights*
Yes (relevant)	1 (not very important)
No (not relevant)	2
Not sure—remind me later	3
	4 (very important)
Answer	*Resilience Score*
No	1 (lowest resilience)
Yes	3 (highest resilience)

#48: To what degree do local versus distant sources influence air quality?

Relevance	*Importance Weights*
Yes (relevant)	1 (not very important)
No (not relevant)	2
Not sure—remind me later	3
	4 (very important)
Answer	*Resilience Score*
Air quality is much more strongly determined by distant sources than local sources and is therefore harder for the urban area to control.	1 (lowest resilience)
Air quality is somewhat more strongly determined by distant sources than local sources and is therefore harder for the urban area to control.	2
Air quality is somewhat more strongly determined by local sources than distant sources and is therefore easier for the urban area to control.	3
Air quality is much more strongly determined by local sources than distant sources and is therefore easier for the urban area to control.	4 (highest resilience)

#49: Does the urban area have air quality districts?

Relevance	Importance Weights
Yes (relevant)	1 (not very important)
No (not relevant)	2
Not sure—remind me later	3
	4 (very important)

Answer	Resilience Score
No	1 (lowest resilience)
Yes	3 (highest resilience)

#50: Has an air quality analysis been completed at multiple scales/resolutions?

Relevance	Importance Weights
Yes (relevant)	1 (not very important)
No (not relevant)	2
Not sure—remind me later	3
	4 (very important)

Answer	Resilience Score
An air quality analysis has not been completed.	1 (lowest resilience)
An air quality analysis has been completed at a one scale/resolution.	2
Air quality analysis has been completed at a few scales/resolutions.	3
Air quality analysis has been completed at many scales/resolutions.	4 (highest resilience)

#51: Does the urban area have health warnings or alerts for days when air quality may be hazardous?

Relevance	*Importance Weights*
Yes (relevant)	1 (not very important)
No (not relevant)	2
Not sure—remind me later	3
	4 (very important)

Answer	*Resilience Score*
No	1 (lowest resilience)
Yes	3 (highest resilience)

#52: Has an analysis of areas with good ventilation (e.g., aligned with prevailing breezes, good tree canopy cover) been completed?

Relevance	*Importance Weights*
Yes (relevant)	1 (not very important)
No (not relevant)	2
Not sure—remind me later	3
	4 (very important)

Answer	*Resilience Score*
An analysis of areas with good ventilation has not been planned or completed.	1 (lowest resilience)
An analysis of areas with good ventilation is planned.	2
An analysis of areas with good ventilation is in progress.	3
An analysis of areas with good ventilation has been completed.	4 (highest resilience)

#53: Do plans exist for preserving areas with good ventilation (e.g., those aligned with prevailing breezes)?

Relevance	*Importance Weights*
Yes (relevant)	1 (not very important)
No (not relevant)	2
Not sure—remind me later	3
	4 (very important)

Answer	*Resilience Score*
No	1 (lowest resilience)
Yes	3 (highest resilience)

#54: Does the urban area have a district-scale (i.e., higher resolution than city scale) thermal comfort index?

Relevance	*Importance Weights*
Yes (relevant)	1 (not very important)
No (not relevant)	2
Not sure—remind me later	3
	4 (very important)

Answer	*Resilience Score*
No	1 (lowest resilience)
Yes	3 (highest resilience)

I.5. People

The questions below have been developed for the people sector. Each question is flagged with one or more of the following gradual change climate stressor and/or extreme event climate stressor (from the urban resilience framework developed for this project):

Stressors

Gradual Changes	*Extreme Events*
Wind speed	Magnitude/duration of heat waves
Temperature	Drought intensity/duration
Precipitation	Flood magnitude/frequency
Sea level rise	Hurricane intensity/frequency
	Storm surge/flooding

In addition, each question has up to four possible **answers**. Each answer has been assigned a **resilience score** on a scale of 1 (lowest resilience) to 4 (highest resilience).

For each question, please:

1. Discuss the **relevance** of the question to the people sector. (If unsure, please select the *not sure—remind me later* option.) Questions may be selected as *yes (relevant)* on the basis of the stressors previously selected as being most relevant to Washington, DC or based on any other criteria.
2. For questions marked as *yes (relevant)*, discuss an **importance weight**, where 1 = not very important and 4 = very important.
3. For questions marked as *yes (relevant)*, identify the best **answer** to the question from the options provided.

#55: How available and how comprehensive are your planning resources for responding to extreme events?

Relevance	Importance Weights
Yes (relevant)	1 (not very important)
No (not relevant)	2
Not sure—remind me later	3
	4 (very important)

Answer	Resilience Score
Comprehensive planning resources for responding to extreme events do not exist or are difficult to access for some of the population.	1 (lowest resilience)
Comprehensive planning resources for responding to extreme events are difficult to access for some of the population.	2
Comprehensive planning resources for responding to extreme events are readily available to some of the population.	3
Comprehensive planning resources for responding to extreme events are readily available to most or all of the population.	4 (highest resilience)

#56: Are government-led, community-based, or other organizations actively promoting adaptive behaviors at the neighborhood or city level?

Relevance	Importance Weights
Yes (relevant)	1 (not very important)
No (not relevant)	2
Not sure—remind me later	3
	4 (very important)

Answer	Resilience Score
No	1 (lowest resilience)
Yes	3 (highest resilience)

#57: Do policies and outreach/education programs promote behavioral changes that facilitate climate change adaptation?

Relevance	*Importance Weights*
Yes (relevant)	1 (not very important)
No (not relevant)	2
Not sure—remind me later	3
	4 (very important)

Answer	*Resilience Score*
No	1 (lowest resilience)
Yes	3 (highest resilience)

#58: Are emergency response staff well trained to respond to large-scale extreme weather events?

Relevance	*Importance Weights*
Yes (relevant)	1 (not very important)
No (not relevant)	2
Not sure—remind me later	3
	4 (very important)

Answer	*Resilience Score*
Training does not include instruction in triage and other procedures, such as coordination, during emergencies that affect large numbers of people.	1 (lowest resilience)
Training includes minimal instruction in triage and other procedures, such as coordination, during emergencies that affect large numbers of people.	2
Yes, training includes some instruction in triage and other procedures, such as coordination, during emergencies that affect large numbers of people.	3
Yes, training includes triage and other procedures, such as coordination, during emergencies that affect large numbers of people.	4 (highest resilience)

#59: Is the distribution of public health workers and emergency response resources appropriate for the population that would be affected during an extreme event?

Relevance	Importance Weights
Yes (relevant)	1 (not very important)
No (not relevant)	2
Not sure—remind me later	3
	4 (very important)

Answer	Resilience Score
The distribution of such services could use improvement.	1 (lowest resilience)
Yes, such services are well-distributed.	3 (highest resilience)

#60: Is there sufficient capacity in public health and emergency response systems for responding to extreme events?

Relevance	Importance Weights
Yes (relevant)	1 (not very important)
No (not relevant)	2
Not sure—remind me later	3
	4 (very important)

Answer	Resilience Score
No	1 (lowest resilience)
Yes	3 (highest resilience)

#61: Does the city have the capacity to provide public transportation for emergency evacuations?

Relevance	Importance Weights
Yes (relevant)	1 (not very important)
No (not relevant)	2
Not sure—remind me later	3
	4 (very important)

Answer	Resilience Score
Insufficient capacity	1 (lowest resilience)
Fair capacity	2
Moderate capacity	3
Extensive capacity	4 (highest resilience)

#62: What evacuation and shelter-in-place options are available to residents in the event of a heat wave?

Relevance	Importance Weights
Yes (relevant)	1 (not very important)
No (not relevant)	2
Not sure—remind me later	3
	4 (very important)

Answer	Resilience Score
No evacuation or shelter-in-place options are available to residents in the event of a heat wave.	1 (lowest resilience)
One to two evacuation and shelter-in-place options are available to residents in the event of a heat wave.	2
Several evacuation and shelter-in-place options are available to residents in the event of a heat wave.	3
Many evacuation and shelter-in-place options are available to residents in the event of a heat wave.	4 (highest resilience)

#63: Do plans exist to provide public access to cooling centers or for other heat adaptation strategies (e.g., opening public swimming pools earlier or later than normal, using fire hydrants for cooling), given predicted climatic changes?

Relevance	*Importance Weights*
Yes (relevant)	1 (not very important)
No (not relevant)	2
Not sure—remind me later	3
	4 (very important)

Answer	*Resilience Score*
Plans do not exist to provide heat adaptation strategies.	1 (lowest resilience)
Plans exist to provide one or a few heat adaptation strategies.	2
Plans exist to provide some heat adaptation strategies.	3
Plans exist to provide many heat adaptation strategies.	4 (highest resilience)

#64: Is the healthcare community, including primary care physicians, prepared for changes in patients' treatments necessitated by climate change (e.g., emerging infectious diseases)?

Relevance	*Importance Weights*
Yes (relevant)	1 (not very important)
No (not relevant)	2
Not sure—remind me later	3
	4 (very important)

Answer	*Resilience Score*
The healthcare community is poorly prepared.	1 (lowest resilience)
The healthcare community's level of preparation is fair.	2
Yes, the healthcare community is moderately prepared.	3
Yes, the healthcare community is well-prepared.	4 (highest resilience)

#65: Is the availability of public health goods and services at risk if other city goods and services (e.g., power, water, public transportation) are affected by extreme climatic events or gradual climatic changes?

Relevance	Importance Weights
Yes (relevant)	1 (not very important)
No (not relevant)	2
Not sure—remind me later	3
	4 (very important)

Answer	Resilience Score
Availability of public health resources is at significant risk if other city services are affected by climatic events or changes.	1 (lowest resilience)
Availability of public health resources is at moderate risk if other city services are affected by climatic events or changes.	2
Availability of public health resources is at some risk if other city services are affected by climatic events or changes.	3
Availability of public health resources is at minimal risk if other city services are affected by climatic events or changes.	4 (highest resilience)

#66: Do public health programs incorporate longer time frames (e.g., 10 or more years), and do they address climate change-related health issues (e.g., movement of deer ticks to more northerly locations)?

Relevance	Importance Weights
Yes (relevant)	1 (not very important)
No (not relevant)	2
Not sure—remind me later	3
	4 (very important)

Answer	Resilience Score
Public health programs are not designed to address climate-related health issues.	1 (lowest resilience)
Public health programs incorporate long-term timeframes and are address climate-related health issues.	3 (highest resilience)

#67: Have public health agencies identified infectious diseases and/or disease vectors that may become more prevalent in the urban area under the expected climatic changes?

Relevance	*Importance Weights*
Yes (relevant)	1 (not very important)
No (not relevant)	2
Not sure—remind me later	3
	4 (very important)

Answer	*Resilience Score*
No	1 (lowest resilience)
Yes	3 (highest resilience)

#68: Have public health agencies developed plans for responding to increased disease and vector exposure in ways that may reduce the associated morbidity/mortality?

Relevance	*Importance Weights*
Yes (relevant)	1 (not very important)
No (not relevant)	2
Not sure—remind me later	3
	4 (very important)

Answer	*Resilience Score*
No	1 (lowest resilience)
Yes	3 (highest resilience)

#69: Do planners in the urban area know the demographic characteristics of populations vulnerable to climate change?

Relevance	*Importance Weights*
Yes (relevant)	1 (not very important)
No (not relevant)	2
Not sure—remind me later	3
	4 (very important)

Answer	*Resilience Score*
No	1 (lowest resilience)
Yes	3 (highest resilience)

#70: Do planners in the urban area know the locations of populations most vulnerable to climate change effects?

Relevance	*Importance Weights*
Yes (relevant)	1 (not very important)
No (not relevant)	2
Not sure—remind me later	3
	4 (very important)

Answer	*Resilience Score*
No	1 (lowest resilience)
Yes	3 (highest resilience)

#71: Are there services and emergency responses aimed at quickly reaching vulnerable populations during power outages?

Relevance	*Importance Weights*
Yes (relevant)	1 (not very important)
No (not relevant)	2
Not sure—remind me later	3
	4 (very important)
Answer	*Resilience Score*
Services and emergency responses are not made especially available to vulnerable populations during power outages.	1 (lowest resilience)
Yes, but these services and responses are provided slower than they are needed.	2
Yes, and these services and responses are provided somewhat rapidly.	3
Yes, and these services and responses are provided rapidly.	4 (highest resilience)

#72: Are policies and programs to promote adaptive behavior designed with frames/messaging that reach the critical audiences in the urban area?

Relevance	*Importance Weights*
Yes (relevant)	1 (not very important)
No (not relevant)	2
Not sure—remind me later	3
	4 (very important)
Answer	*Resilience Score*
No	1 (lowest resilience)
Yes	3 (highest resilience)

#73: Are policies and programs to promote adaptive behavior designed and implemented in ways that promote the health and well-being of vulnerable populations?

Relevance	Importance Weights
Yes (relevant)	1 (not very important)
No (not relevant)	2
Not sure—remind me later	3
	4 (very important)

Answer	Resilience Score
No	1 (lowest resilience)
Yes	3 (highest resilience)

#74: Are policies and programs to promote adaptive behavior evaluated in ways that take into account vulnerable populations?

Relevance	Importance Weights
Yes (relevant)	1 (not very important)
No (not relevant)	2
Not sure—remind me later	3
	4 (very important)

Answer	Resilience Score
No	1 (lowest resilience)
Yes	3 (highest resilience)

#108: How accessible are different modes of transportation (e.g., to what proportion of the population, what subpopulations [vulnerable people])?

Relevance	Importance Weights
Yes (relevant)	1 (not very important)
No (not relevant)	2
Not sure—remind me later	3
	4 (very important)

Answer	Resilience Score
Few to no modes of transportation are accessible to vulnerable subpopulations.	1 (lowest resilience)
Some modes of transportation are accessible to vulnerable subpopulations.	2
Many modes of transportation are accessible to vulnerable subpopulations.	3
All modes of transportation are accessible to vulnerable subpopulations.	4 (highest resilience)

#109: What proportion of the population has limited access to transportation options due to compromised health or lower income levels? For what proportion of this population might transportation failures be life-threatening (e.g., due to reduced access to specialized medical care or equipment)?

Relevance	*Importance Weight*
Yes (relevant)	1 (not very important)
No (not relevant)	2
Not sure—remind me later	3
	4 (very important)

Answer	*Resilience Score*
10% or more of the population has limited access to transportation due to vulnerabilities, and transportation failures may be life-threatening for *25% or more* of this population.	1 (lowest resilience)
10% or more of the population has limited access to transportation due to vulnerabilities, and transportation failures may be life-threatening for *less than 25%* of this population.	2
Less than 10% of the population has limited access to transportation due to vulnerabilities, and transportation failures may be life-threatening for *25% or more* of this population.	3
Less than 10% of the population has limited access to transportation due to vulnerabilities, and transportation failures may be life-threatening for *less than 25%* of this population.	4 (highest resilience)

#143: Are early warning systems for meteorological extreme events available?

Relevance	Importance Weight
Yes (relevant)	1 (not very important)
No (not relevant)	2
Not sure—remind me later	3
	4 (very important)

Answer	Resilience Score
No	1 (lowest resilience)
Yes	3 (highest resilience)

#158: Do municipal managers consider local stakeholder knowledge and local resources (e.g., libraries, archives) in climate change resilience planning?

Relevance	Importance Weight
Yes (relevant)	1 (not very important)
No (not relevant)	2
	3
	4 (very important)

Answer	Resilience Score
No	1 (lowest resilience)
Yes	3 (highest resilience)

I.6. Telecommunications

The questions below have been developed for the telecommunication sector. Each question is flagged with one or more of the following gradual change climate stressor and/or extreme event climate stressor (from the urban resilience framework developed for this project):

Stressors

Gradual Changes	*Extreme Events*
Wind speed	Magnitude/duration of heat waves
Temperature	Drought intensity/duration
Precipitation	Flood magnitude/frequency
Sea level rise	Hurricane intensity/frequency
	Storm surge/flooding

In addition, each question has up to four possible **answers**. Each answer has been assigned a **resilience score** on a scale of 1 (lowest resilience) to 4 (highest resilience).

For each question, please:

1. Discuss the **relevance** of the question to the telecommunication sector. (If unsure, please select the *not sure—remind me later* option.) Questions may be selected as *yes (relevant)* on the basis of the stressors previously selected as being most relevant to Washington, DC or based on any other criteria.
2. For questions marked as *yes (relevant)*, discuss an **importance weight**, where 1 = not very important and 4 = very important.

For questions marked as *yes (relevant)*, identify the best **answer** to the question from the options provided.

#75: What natural disasters has the area experienced in the past, and what services were retained or largely unaffected despite these disasters?

Relevance	*Importance Weights*
Yes (relevant)	1 (not very important)
No (not relevant)	2
Not sure—remind me later	3
	4 (very important)
Answer	*Resilience Score*
Area has either not experienced many natural disasters in recent history, or services were significantly impaired during recent natural disasters.	1 (lowest resilience)
Area has experienced some extreme weather or other natural disasters, but some services were significantly affected.	2
Area has experienced some extreme weather or other natural disasters, and most services were unaffected or affected in minor ways.	3
Area has experienced major extreme weather events or other natural disasters, and majority of services were retained or were largely unaffected.	4 (highest resilience)

#76: How would a temporary loss of telecommunication infrastructure affect the local and regional economies?

Relevance	*Importance Weights*
Yes (relevant)	1 (not very important)
No (not relevant)	2
Not sure—remind me later	3
	4 (very important)
Answer	*Resilience Score*
Major effect	1 (lowest resilience)
Moderate effect	2
Small effect	3
Little to no effect	4 (highest resilience)

#77: Are data centers located within or outside of the urban area?

Relevance	Importance Weights
Yes (relevant)	1 (not very important)
No (not relevant)	2
Not sure—remind me later	3
	4 (very important)

Answer	Resilience Score
Within	1 (lowest resilience)
Mostly within the urban area, but somewhat outside the urban area.	2
Mostly outside the urban area, but somewhat within the urban area.	3
Outside	4 (highest resilience)

#78: For each telecommunication service, are there key nodes whose failure would severely affect the service?

Relevance	Importance Weight
Yes (relevant)	1 (not very important
No (not relevant)	2
Not sure—remind me later	3
	4 (very important)

Answer	Resilience Score
There are many key nodes whose failure would severely affect service.	1 (lowest resilience)
There are some key nodes whose failure would severely affect service.	2
There are a few key nodes whose failure would severely affect service.	3
No, there are no nodes whose failure would severely affect service.	4 (highest resilience)

#79: How robust is the telecommunication network in terms of resilience to damage to or failure of key nodes?

Relevance	*Importance Weights*
Yes (relevant)	1 (not very important)
No (not relevant)	2
Not sure—remind me later	3
	4 (very important)

Answer	*Resilience Score*
The telecommunication network is not resilient to damage or failure of key nodes.	1 (lowest resilience)
The telecommunication network is slightly resilient to damage or failure of key nodes.	2
The telecommunication network is somewhat resilient to damage or failure of key nodes.	3
The telecommunication network is very resilient to damage or failure of key nodes.	4 (highest resilience)

#80: Are there parts of the telecommunication infrastructure that are particularly vulnerable to high temperatures or prolonged high temperatures?

Relevance	*Importance Weights*
Yes (relevant)	1 (not very important)
No (not relevant)	2
Not sure—remind me later	3
	4 (very important)

Answer	*Resilience Score*
No	1 (lowest resilience)
Yes	3 (highest resilience)

#81: Are there satellite-based communications on frequency bands (e.g., the Ka band) that are vulnerable to wet-weather disruption?

Relevance	Importance Weights
Yes (relevant)	1 (not very important)
No (not relevant)	2
Not sure—remind me later	3
	4 (very important)

Answer	Resilience Score
No	1 (lowest resilience)
Yes	3 (highest resilience)

#82: Are your telecommunication infrastructure components located wisely with respect to your anticipated climate stressors (i.e., aboveground, underground, or serviced by satellite)?

Relevance	Importance Weights
Yes (relevant)	1 (not very important)
No (not relevant)	2
Not sure—remind me later	3
	4 (very important)

Answer	Resilience Score
No	1 (lowest resilience)
Yes	3 (highest resilience)

#83: Are aboveground infrastructure components vulnerable to wind (e.g., cell towers)?

Relevance	Importance Weights
Yes (relevant)	1 (not very important)
No (not relevant)	2
Not sure—remind me later	3
	4 (very important)

Answer	Resilience Score
All aboveground infrastructure components are vulnerable to expected winds.	1 (lowest resilience)
Some aboveground infrastructure components are vulnerable to expected winds.	2
Few aboveground infrastructure components are vulnerable to expected winds.	3
No aboveground infrastructure components are vulnerable to expected winds.	4 (highest resilience)

#84: Are belowground infrastructure components vulnerable to rising water or salt water intrusion?

Relevance	Importance Weights
Yes (relevant)	1 (not very important)
No (not relevant)	2
Not sure—remind me later	3
	4 (very important)

Answer	Resilience Score
All belowground infrastructure components are vulnerable to expected rises in groundwater levels or from salt water intrusion.	1 (lowest resilience)
Some belowground infrastructure components are vulnerable to expected rises in groundwater levels or from salt water intrusion.	2
Few belowground infrastructure components are vulnerable to expected rises in groundwater levels or from salt water intrusion.	3
No belowground infrastructure components are vulnerable to expected rises in groundwater levels or from salt water intrusion.	4 (highest resilience)

#85: If the area has satellite-based communications that are vulnerable to wet-weather disruption, does the area have a backup tower network?

Relevance	*Importance Weight*
Yes (relevant)	1 (not very important)
No (not relevant)	2
Not sure—remind me later	3
	4 (very important)

Answer	*Resilience Score*
The area does not have a tower network that could provide backup.	1 (lowest resilience)
The area has a tower network that could provide a small amount of backup.	2
The area has a tower network that could provide some backup.	3
The area has a tower network that could provide full backup to satellite-based communications.	4 (highest resilience)

#86: Does your community have sufficient access to backup telecommunication systems? What is the capacity of the telecommunication infrastructure?

Relevance	*Importance Weights*
Yes (relevant)	1 (not very important)
No (not relevant)	2
Not sure—remind me later	3
	4 (very important)

Answer	*Resilience Score*
There are no backup systems. Capacity of the telecommunication infrastructure is low.	1 (lowest resilience)
There are some minimal backup systems, but telecommunication infrastructure capacity is likely to be a problem during an emergency.	2
There are some backup systems in place. Capacity of the systems is moderate.	3
Backup systems are in place. Capacity of the telecommunication systems is high.	4 (highest resilience)

#87: Is backup power for telecommunication systems provided? If so, is it provided by diesel generators?

Relevance	Importance Weights
Yes (relevant)	1 (not very important)
No (not relevant)	2
Not sure—remind me later	3
	4 (very important)

Answer	Resilience Score
Backup power is not provided.	1 (lowest resilience)
Backup power is provided, but it is provided by diesel generators.	2
Backup power is provided and is only partially provided by diesel generators.	3
Backup power is provided and is not provided by diesel generators.	4 (highest resilience)

#88: What is the extent of telecommunication redundancy? Do first responders and the public have multiple communication options, served by different infrastructure?

Relevance	Importance Weights
Yes (relevant)	1 (not very important)
No (not relevant)	2
Not sure—remind me later	3
	4 (very important)

Answer	Resilience Score
There is little to no redundancy.	1 (lowest resilience)
There is a small amount of redundancy.	2
There is a moderate amount of redundancy. There are more than one communications options, served by different infrastructure.	3
There is a great deal of redundancy. There are multiple communication options, served by different infrastructure.	4 (highest resilience)

#89: What percentage of telecommunication system capacity is required for the baseline level of use?

Relevance	Importance Weight
Yes (relevant)	1 (not very important)
No (not relevant)	2
Not sure—remind me later	3
	4 (very important)

Answer	Resilience Score
Greater than 85%	1 (lowest resilience)
70 to 85%	2
60 to 70%	3
Less than 60%	4 (highest resilience)

#90: Does telecommunication infrastructure have the capacity for increased public demand in an emergency?

Relevance	Importance Weights
Yes (relevant)	1 (not very important)
No (not relevant)	2
Not sure—remind me later	3
	4 (very important)

Answer	Resilience Score
No	1 (lowest resilience)
Yes	3 (highest resilience)

#91: Do local authorities have established relations with telecommunication infrastructure service providers? Are emergency protocols and plans in place?

Relevance	Importance Weights
Yes (relevant)	1 (not very important)
No (not relevant)	2
Not sure—remind me later	3
	4 (very important)

Answer	Resilience Score
No	1 (lowest resilience)
Yes	3 (highest resilience)

#92: Do local private-sector telecommunication infrastructure service providers have the authority and resources to make quick decisions and implement them in and after an emergency?

Relevance	Importance Weights
Yes (relevant)	1 (not very important)
No (not relevant)	2
Not sure—remind me later	3
	4 (very important)

Answer	Resilience Score
No	1 (lowest resilience)
Yes	3 (highest resilience)

#93: Can local authorities and telecommunication providers give first responder and decision-maker communications priority during an expected surge in traffic in emergency situations?

Relevance	*Importance Weights*
Yes (relevant)	1 (not very important)
No (not relevant)	2
Not sure—remind me later	3
	4 (very important)

Answer	*Resilience Score*
No	1 (lowest resilience)
Yes	3 (highest resilience)

#94: Are public-address systems (e.g., loud speakers, text messages, radio broadcasts, emergency television broadcasts) in place to provide instructions to the public in case of an emergency?

Relevance	*Importance Weight*
Yes (relevant)	1 (not very important)
No (not relevant)	2
Not sure—remind me later	3
	4 (very important)

Answer	*Resilience Score*
There are no public-address systems in place.	1 (lowest resilience)
There are insufficient public-address systems in place.	2
Some public-address systems are in place, but there could be more.	3
Sufficient public-address systems are in place.	4 (highest resilience)

#95: What modes do authorities in the urban area use to communicate emergency information and alerts? Are these modes low or high bandwidth?

Relevance	Importance Weights
Yes (relevant)	1 (not very important)
No (not relevant)	2
Not sure—remind me later	3
	4 (very important)

Answer	Resilience Score
Authorities do not use multiple modes (e.g., text messaging, email, phone calls), or none of the modes used is low bandwidth.	1 (lowest resilience)
Authorities use one to two modes (e.g., text messaging, email, phone calls) and one or two of these modes is low bandwidth.	2
Authorities use multiple modes (e.g., text messaging, email, phone calls) and one or two of these modes are low bandwidth.	3
Authorities use multiple modes (e.g., text messaging, email, phone calls) and some of these modes are low bandwidth.	4 (highest resilience)

#96: What is the likelihood that the capacity of local first responder communication systems would be exceeded during a disaster?

Relevance	*Importance Weights*
Yes (relevant)	1 (not very important)
No (not relevant)	2
Not sure—remind me later	3
	4 (very important)

Answer	*Resilience Score*
It is very likely that the capacity of local first responder communications would be exceeded during a disaster.	1 (lowest resilience)
It is somewhat likely that the capacity of local first responder communications would be exceeded during a disaster.	2
It is somewhat unlikely that the capacity of local first responder communications would be exceeded during a disaster.	3
It is very unlikely that the capacity of local first responder communications would be exceeded during a disaster.	4 (highest resilience)

#97: Does the area have access to backup emergency call/response (911) networks if the primary networks fail or are overloaded?

Relevance	*Importance Weights*
Yes (relevant)	1 (not very important)
No (not relevant)	2
Not sure—remind me later	3
	4 (very important)

Answer	*Resilience Score*
No, or the backup network could handle only a minimal amount of the load for the main emergency response network.	1 (lowest resilience)
Yes, but the backup network could handle only some of the load for the main emergency response network.	2
Yes, and the backup network could handle the most of the load for the main emergency response network.	3
Yes, and the backup network could handle the entire load for the main emergency response network.	4 (highest resilience)

#98: Is the availability of telecommunication goods and services at risk if other city goods and services (e.g., power, water, transportation) are affected by extreme climatic events or gradual climatic changes?

Relevance	Importance Weights
Yes (relevant)	1 (not very important)
No (not relevant)	2
Not sure—remind me later	3
	4 (very important)

Answer	Resilience Score
Availability of telecommunication resources is at significant risk if other city services are affected by climatic events or changes.	1 (lowest resilience)
Availability of telecommunication resources is at moderate risk if other city services are affected by climatic events or changes.	2
Availability of telecommunication resources is at some risk if other city services are affected by climatic events or changes.	3
Availability of telecommunication resources is at minimal risk if other city services are affected by climatic events or changes.	4 (highest resilience)

#99: Do telecommunication systems have enough energy and water supply to handle an extra load in the case of sudden natural disasters?

Relevance	*Importance Weight*
Yes (relevant)	1 (not very important)
No (not relevant)	2
Not sure—remind me later	3
	4 (very important)

Answer	*Resilience Score*
Systems do not have enough to handle any of the anticipated extra load.	1 (lowest resilience)
Systems have enough to handle a small amount of the anticipated extra load.	2
Systems have enough to handle some of the anticipated extra load.	3
Systems have enough to handle all of the anticipated extra load.	4 (highest resilience)

#160: Have city planners consulted with other city governments with similar telecommunication systems to learn from their experience with natural disasters and prepare for similar events?

Relevance	*Importance Weight*
Yes (relevant)	1 (not very important)
No (not relevant)	2
Not sure—remind me later	3
	4 (very important)

Answer	*Resilience Score*
No	1 (lowest resilience)
Yes	3 (highest resilience)

I.7. Transportation

The questions below have been developed for the transportation sector. Each question is flagged with one or more of the following gradual change climate stressor and/or extreme event climate stressor (from the urban resilience framework developed for this project):

Stressors

Gradual Changes
- Wind speed
- Temperature
- Precipitation
- Sea level rise

Extreme Events
- Magnitude/duration of heat waves
- Drought intensity/duration
- Flood magnitude/frequency
- Hurricane intensity/frequency
- Storm surge/flooding

In addition, each question has up to four possible **answers**. Each answer has been assigned a **resilience score** on a scale of 1 (lowest resilience) to 4 (highest resilience).

For each question, please:

1. Discuss the **relevance** of the question to the transportation sector. (If unsure, please select the *not sure—remind me later* option.) Questions may be selected as *yes (relevant)* on the basis of the stressors previously selected as being most relevant to Washington, DC or based on any other criteria.
2. For questions marked as *yes (relevant)*, discuss an **importance weight**, where 1 = not very important and 4 = very important.
3. For questions marked as *yes (relevant)*, identify the best **answer** to the question from the options provided.

#100: Is the availability of transportation goods and services at risk if other city goods and services (e.g., power, water, telecommunications) are affected by extreme climatic events or gradual climatic changes?

Relevance	*Importance Weights*
Yes (relevant)	1 (not very important)
No (not relevant)	2
Not sure—remind me later	3
	4 (very important)
Answer	*Resilience Score*
Availability of transportation resources is at significant risk if other city services are affected by climatic events or changes.	1 (lowest resilience)
Availability of transportation resources is at moderate risk if other city services are affected by climatic events or changes.	2
Availability of transportation resources is at some risk if other city services are affected by climatic events or changes.	3
Availability of transportation resources is at minimal risk if other city services are affected by climatic events or changes.	4 (highest resilience)

#101: How much risk is assumed in the design of transportation systems (bridges, culverts), and does it span the anticipated changes in precipitation, temperature, and storm intensities under climate change?

Relevance	*Importance Weights*
Yes (relevant)	1 (not very important)
No (not relevant)	2
Not sure—remind me later	3
	4 (very important)
Answer	*Resilience Score*
None	1 (lowest resilience)
Low	2
Medium	3
High	4 (highest resilience)

#102: How resistant to potential impacts of climate change are critical transportation facilities (e.g., high-traffic vehicle or rail bridges, tunnels)?

Relevance	*Importance Weights*
Yes (relevant)	1 (not very important)
No (not relevant)	2
Not sure—remind me later	3
	4 (very important)

Answer	*Resilience Score*
Critical transportation facilities are not at all resistant or have no redundancy.	1 (lowest resilience)
Critical transportation facilities are not very resistant or have low levels of redundancy.	2
Critical transportation facilities are moderately resistant or have moderate levels of redundancy.	3
Critical transportation facilities are very resistant or have high levels of redundancy.	4 (highest resilience)

#103: What degree of redundancy exists for major transportation links? Are there single points of failure? What are the implications of losing a particular link, and how rapidly can you recover?

Relevance	*Importance Weights*
Yes (relevant)	1 (not very important)
No (not relevant)	2
Not sure—remind me later	3
	4 (very important)

Answer	*Resilience Score*
Little to no redundancy exists for most links, so there is a single point of failure in transportation systems and recovery would be slow.	1 (lowest resilience)
Some redundancy exists for most links, so few systems have single points of failure, but recovery would be slow.	2
Some redundancy exists for most links, so few systems have single points of failure and recovery would be rapid.	3
Significant redundancy exists for most links, so few to no systems have single points of failure, and recovery would be rapid.	4 (highest resilience)

#104: What length of time would be required to restore major high-traffic vehicle transportation links in the urban area if they experience a failure?

Relevance	Importance Weights
Yes (relevant)	1 (not very important)
No (not relevant)	2
Not sure—remind me later	3
	4 (very important)

Answer	Resilience Score
More than 1 week	1 (lowest resilience)
Approximately 1 week	2
4 to 6 days	3
1 to 3 days	4 (highest resilience)

#105: Are any portions of the transportation system less important if the duration of the disturbance is a few days? What if the duration of the disturbance is more on the order of weeks?

Relevance	Importance Weights
Yes (relevant)	1 (not very important)
No (not relevant)	2
Not sure—remind me later	3
	4 (very important)

Answer	Resilience Score
No, all components of the transportation system are critical to the functioning of transportation in the area.	1 (lowest resilience)
A few portions of the transportation system are less important if the disturbance is a few days but not if the disturbance is a few weeks.	2
Several portions of the transportation system are less important if the disturbance is a few days but not if the disturbance is a few weeks.	3
Some portions of the transportation system are less important whether the disturbance is a few days or a few weeks.	4 (highest resilience)

#106: To what extent is the area dependent on long-range transportation of goods and services versus locally available goods and services (food, energy, etc.)?

Relevance	Importance Weights
Yes (relevant)	1 (not very important)
No (not relevant)	2
Not sure—remind me later	3
	4 (very important)
Answer	Resilience Score
90–100% dependent on long-range transportation of goods and services	1 (lowest resilience)
50–90% dependent on long-range transportation of goods and services	2
10–50% dependent on long-range transportation of goods and services	3
0–10% dependent on long-range transportation of goods and services	4 (highest resilience)

#107: What flexibility has been built into the transportation system (different modes)?

Relevance	Importance Weights
Yes (relevant)	1 (not very important)
No (not relevant)	2
Not sure—remind me later	3
	4 (very important)
Answer	Resilience Score
1–2 modes available	1 (lowest resilience)
3–4 modes available	2
5–6 modes available	3
7 or more modes available	4 (highest resilience)

#108: How accessible are different modes (e.g., to what proportion of the population, what subpopulations [vulnerable people])?

Relevance	*Importance Weights*
Yes (relevant)	1 (not very important)
No (not relevant)	2
Not sure—remind me later	3
	4 (very important)

Answer	*Resilience Score*
Few to no modes of transportation are accessible to vulnerable subpopulations.	1 (lowest resilience)
Some modes of transportation are accessible to vulnerable subpopulations.	2
Many modes of transportation are accessible to vulnerable subpopulations.	3
All modes of transportation are accessible to vulnerable subpopulations.	4 (highest resilience)

#109: What proportion of the population has limited access to transportation options due to compromised health or lower income levels? For what proportion of this population might transportation failures be life-threatening (e.g., due to reduced access to specialized medical care or equipment)?

Relevance	*Importance Weights*
Yes (relevant)	1 (not very important)
No (not relevant)	2
Not sure—remind me later	3
	4 (very important)

Answer	*Resilience Score*
10% or more of the population has limited access to transportation due to vulnerabilities, and transportation failures may be life-threatening for *25% or more* of this population.	1 (lowest resilience)
10% or more of the population has limited access to transportation due to vulnerabilities, and transportation failures may be life-threatening for *less than 25%* of this population.	2
Less than 10% of the population has limited access to transportation due to vulnerabilities, and transportation failures may be life-threatening for *25% or more* of this population.	3
Less than 10% of the population has limited access to transportation due to vulnerabilities, and transportation failures may be life-threatening for *less than 25%* of this population.	4 (highest resilience)

#110: How familiar is the community with evacuation procedures?

Relevance	*Importance Weights*
Yes (relevant)	1 (not very important)
No (not relevant)	2
Not sure—remind me later	3
	4 (very important)

Answer	*Resilience Score*
Unfamiliar	1 (lowest resilience)
Only slightly familiar (or only some subpopulations are familiar)	2
Somewhat familiar	3
Very familiar	4 (highest resilience)

#111: What length of time would be required to restore major passenger rail transportation facilities in the urban area if they experience a failure?

Relevance	Importance Weights
Yes (relevant)	1 (not very important)
No (not relevant)	2
Not sure—remind me later	3
	4 (very important)

Answer	Resilience Score
More than 1 week	1 (lowest resilience)
Approximately 1 week	2
4 to 6 days	3
1 to 3 days	4 (highest resilience)

#112: What length of time would be required to restore major freight rail transportation facilities in the urban area if they experience a failure?

Relevance	Importance Weights
Yes (relevant)	1 (not very important)
No (not relevant)	2
Not sure—remind me later	3
	4 (very important)

Answer	Resilience Score
More than 1 week	1 (lowest resilience)
Approximately 1 week	2
4 to 6 days	3
1 to 3 days	4 (highest resilience)

#113: What length of time would be required to restore major bicycle and pedestrian transportation links in the urban area if they experience a failure?

Relevance	*Importance Weights*
Yes (relevant)	1 (not very important)
No (not relevant)	2
Not sure—remind me later	3
	4 (very important)

Answer	*Resilience Score*
Approximately 1 week	1 (lowest resilience)
4 to 6 days	2
1 to 3 days	3
Less than 1 day	4 (highest resilience)

#114: Are urban areas set up to provide accessibility (e.g., to jobs) if mobility is interrupted or impeded?

Relevance	*Importance Weights*
Yes (relevant)	1 (not very important)
No (not relevant)	2
Not sure—remind me later	3
	4 (very important)

Answer	*Resilience Score*
No	1 (lowest resilience)
Yes	3 (highest resilience)

#115: Do current planning regimes include proactive resilience building, or is only reactive disaster response being addressed?

Relevance	*Importance Weights*
Yes (relevant)	1 (not very important)
No (not relevant)	2
Not sure—remind me later	3
	4 (very important)

Answer	*Resilience Score*
Current planning regime only addresses reactive disaster response.	1 (lowest resilience)
Current planning regime only addresses reactive disaster response, but proactive resilience-building approaches are being developed.	2
Proactive resilience-building approaches have been developed and are being implemented alongside reactive disaster response plans.	3
Proactive resilience-building approaches are implemented alongside reactive disaster response plans.	4 (highest resilience)

#116: Are there funding mechanisms that exist or could be put into place to complete the necessary work on the transportation system to adapt to anticipated climatic changes and increased risks?

Relevance	*Importance Weights*
Yes (relevant)	1 (not very important)
No (not relevant)	2
Not sure—remind me later	3
	4 (very important)

Answer	*Resilience Score*
No funding mechanisms exist to adapt transportation systems to climatic changes, and none could be established.	1 (lowest resilience)
No funding mechanisms exist to adapt transportation systems to climatic changes, but mechanisms could be established.	2
Funding mechanisms are being developed to adapt transportation systems to climatic changes.	3
Funding mechanisms exist to adapt transportation systems to climatic changes.	4 (highest resilience)

#117: Do plans exist to replace aging infrastructure? If so, do these plans account for the anticipated impacts of climate change on this infrastructure?

Relevance	*Importance Weights*
Yes (relevant)	1 (not very important)
No (not relevant)	2
Not sure—remind me later	3
	4 (very important)
Answer	*Resilience Score*
No plans exist to replace aging infrastructure.	1 (lowest resilience)
Plans are being developed or already exist to replace aging infrastructure, but they do not account for anticipated impacts of climate change.	2
Plans are being developed or already exist to replace aging infrastructure, but only some of these plans account for anticipated impacts of climate change.	3
Plans exist to replace aging infrastructure, and these plans account for anticipated impacts of climate change.	4 (highest resilience)

#118: Are the materials currently in use in transportation systems, such as the common asphalt formulations and rail types, compatible with anticipated changes in temperature?

Relevance	*Importance Weights*
Yes (relevant)	1 (not very important)
No (not relevant)	2
Not sure—remind me later	3
	4 (very important)
Answer	*Resilience Score*
No currently used materials are compatible with anticipated changes in temperature.	1 (lowest resilience)
A few currently used materials are compatible with anticipated changes in temperature.	2
Some currently used materials are compatible with anticipated changes in temperature.	3
All currently used materials are compatible with anticipated changes in temperature.	4 (highest resilience)

#119: Have new or innovative materials been tested that may be more capable of withstanding the anticipated impacts of climate change (e.g., higher temperatures)?

Relevance	Importance Weights
Yes (relevant)	1 (not very important)
No (not relevant)	2
Not sure—remind me later	3
	4 (very important)

Answer	Resilience Score
No	1 (lowest resilience)
Yes	3 (highest resilience)

#120: To what extent is green infrastructure implemented or planned to reduce climate change impacts on transportation systems?

Relevance	Importance Weights
Yes (relevant)	1 (not very important)
No (not relevant)	2
Not sure—remind me later	3
	4 (very important)

Answer	Resilience Score
Not implemented or planned	1 (lowest resilience)
Planned but not yet implemented	2
Some implementation with further green infrastructure planned	3
Widespread implementation with additional projects planned	4 (highest resilience)

#162: Have municipalities considered new methods of designing roads/bridges to prepare for heavily traveled routes during an extreme climate event (e.g., coastal evacuation routes)?

Relevance	*Importance Weight*
Yes (relevant)	1 (not very important)
No (not relevant)	2
Not sure—remind me later	3
	4 (very important)

Answer	*Resilience Score*
No	1 (lowest resilience)
Yes	3 (highest resilience)

#168: How resistant to potential impacts of climate change are critical non-road transportation facilities (e.g., high-traffic rail bridges, tunnels)?

Relevance	*Importance Weights*
Yes (relevant)	1 (not very important)
No (not relevant)	2
Not sure—remind me later	3
	4 (very important)

Answer	*Resilience Score*
Critical non-road transportation facilities are not at all resistant or have non redundancy.	1 (lowest resilience)
Critical non-road transportation facilities are not very resistant or have low levels of redundancy.	2
Critical non-road transportation facilities are moderately resistant or have moderate levels of redundancy.	3
Critical non-road transportation facilities are very resistant or have high levels of redundancy.	4 (highest resilience)

#169: Do plans exist to replace aging infrastructure? If so, do these plans account for the anticipated impacts of climate change on this infrastructure?

Relevance	Importance Weights
Yes (relevant)	1 (not very important)
No (not relevant)	2
Not sure—remind me later	3
	4 (very important)

Answer	Resilience Score
No plans exit to replace aging infrastructure.	1 (lowest resilience)
Plans are being developed or already exist to replace aging infrastructure, but they do not account for anticipated impacts of climate change.	2
Plans are being developed or already exist to replace aging infrastructure, but only some of these plans account for anticipated impacts of climate change.	3
Plans exist to replace aging infrastructure and these plans account for anticipated impacts of climate change.	4 (highest resilience)

I.8. Water

The questions below have been developed for the water sector. Each question is flagged with one or more of the following gradual change climate stressor and/or extreme event climate stressor (from the urban resilience framework developed for this project):

Stressors

Gradual Changes	*Extreme Events*
Wind speed	Magnitude/duration of heat waves
Temperature	Drought intensity/duration
Precipitation	Flood magnitude/frequency
Sea level rise	Hurricane intensity/frequency
	Storm surge/flooding

In addition, each question has up to four possible **answers**. Each answer has been assigned a **resilience score** on a scale of 1 (lowest resilience) to 4 (highest resilience).

For each question, please:

1. Discuss the **relevance** of the question to the water sector. (If unsure, please select the *not sure—remind me later* option.) Questions may be selected as *yes (relevant)* on the basis of the stressors previously selected as being most relevant to Washington, DC or based on any other criteria.
2. For questions marked as *yes (relevant)*, discuss an **importance weight**, where 1 = not very important and 4 = very important.
3. For questions marked as *yes (relevant)*, identify the best **answer** to the question from the options provided.

#121: Does the water supply draw from a diversity of sources?

Relevance	Importance Weights
Yes (relevant)	1 (not very important)
No (not relevant)	2
Not sure—remind me later	3
	4 (very important)

Answer	Resilience Score
No	1 (lowest resilience)
Yes	3 (highest resilience)

#122: To what extent do water supplies come from outside the metropolitan area?

Relevance	Importance Weights
Yes (relevant)	1 (not very important)
No (not relevant)	2
Not sure—remind me later	3
	4 (very important)

Answer	Resilience Score
They come exclusively from outside the area.	1 (lowest resilience)
To a great extent	2
To a moderate extent	3
Only to a small extent	4 (highest resilience)

#123: Is there a recharge plan in place for groundwater supplies?

Relevance	Importance Weights
Yes (relevant)	1 (not very important)
No (not relevant)	2
Not sure—remind me later	3
	4 (very important)

Answer	Resilience Score
No	1 (lowest resilience)
Yes	3 (highest resilience)

#124: Do programs for long-term maintenance of water supplies (e.g., erosion control methods, reforestation of the watershed) exist?

Relevance	Importance Weights
Yes (relevant)	1 (not very important)
No (not relevant)	2
Not sure—remind me later	3
	4 (very important)

Answer	Resilience Score
No	1 (lowest resilience)
Yes	3 (highest resilience)

#125: Is there a hierarchy of water uses to be implemented during a shortage or emergency?

Relevance	Importance Weights
Yes (relevant)	1 (not very important)
No (not relevant)	2
Not sure—remind me later	3
	4 (very important)

Answer	Resilience Score
No	1 (lowest resilience)
Yes	3 (highest resilience)

#126: Does the water system have emergency interconnections with adjacent water systems or other emergency sources of supply?

Relevance	Importance Weights
Yes (relevant)	1 (not very important)
No (not relevant)	2
Not sure—remind me later	3
	4 (very important)

Answer	Resilience Score
No	1 (lowest resilience)
Yes	3 (highest resilience)

#127: Are water and wastewater treatment plants located in a flood zone?

Relevance	*Importance Weights*
Yes (relevant)	1 (not very important)
No (not relevant)	2
Not sure—remind me later	3
	4 (very important)

Answer	*Resilience Score*
At least 50% of water and wastewater treatment plant capacity is located in a flood zone.	1 (lowest resilience)
30% to 49% of water and wastewater treatment plant capacity is located in a flood zone.	2
10% to 29% of water and wastewater treatment plant capacity is located in a flood zone.	3
Less than 10% of water and wastewater treatment plant capacity is located in a flood zone.	4 (highest resilience)

#128: Are groundwater supplies susceptible to salt water intrusion and sea level rise?

Relevance	*Importance Weights*
Yes (relevant)	1 (not very important)
No (not relevant)	2
Not sure—remind me later	3
	4 (very important)

Answer	*Resilience Score*
Groundwater supplies are very susceptible to salt water intrusion given anticipated sea level rise.	1 (lowest resilience)
Groundwater supplies are moderately susceptible to salt water intrusion given anticipated sea level rise.	2
Groundwater supplies are slightly susceptible to salt water intrusion given anticipated sea level rise.	3
No, groundwater supplies are not susceptible to salt water intrusion and sea level rise.	4 (highest resilience)

#129: If groundwater supplies are susceptible to salt water intrusion and sea level rise, is the water treatment plant equipped to deal with higher levels of salinity?

Relevance	Importance Weights
Yes (relevant)	1 (not very important)
No (not relevant)	2
Not sure—remind me later	3
	4 (very important)

Answer	Resilience Score
No	1 (lowest resilience)
Yes	3 (highest resilience)

#130: Does treatment capacity exist to accommodate nutrient loading?

Relevance	Importance Weights
Yes (relevant)	1 (not very important)
No (not relevant)	2
Not sure—remind me later	3
	4 (very important)

Answer	Resilience Score
Drinking water treatment capacity cannot accommodate nutrient loading in source water.	1 (lowest resilience)
Drinking water treatment capacity can accommodate expected levels of nutrient loading in source water.	3 (highest resilience)

#131: Does the drinking water treatment plant have redundant treatment chemical suppliers?

Relevance	Importance Weight
Yes (relevant	1 (not very important
No (not relevant)	2
Not sure—remind me later	3
	4 (very important)

Answer	Resilience Score
No	1 (lowest resilience)
Yes	3 (highest resilience)

#132: Are there redundant drinking water systems in place for coping with extreme events, including supply, treatment, and distribution systems?

Relevance	Importance Weights
Yes (relevant)	1 (not very important)
No (not relevant)	2
Not sure—remind me later	3
	4 (very important)

Answer	Resilience Score
No, redundant drinking water systems are not in place.	1 (lowest resilience)
Yes, but these redundant drinking water systems have only a small amount of the capacity necessary.	2
Yes, and these redundant drinking water systems have some of the capacity necessary.	3
Yes, and these redundant drinking water systems have all the capacity necessary.	4 (highest resilience)

#133: Is backup power for water supply, treatment, and distribution systems provided?

Relevance	Importance Weights
Yes (relevant)	1 (not very important)
No (not relevant)	2
Not sure—remind me later	3
	4 (very important)

Answer	Resilience Score
No backup power is provided.	1 (lowest resilience)
Minimal backup power is provided.	2
Some backup power is provided.	3
Full backup power is provided.	4 (highest resilience)

#134: How diverse are individual properties (i.e., are they equipped to harvest rainwater or recharge groundwater so they can create or augment local water supplies)?

Relevance	*Importance Weights*
Yes (relevant)	1 (not very important)
No (not relevant)	2
Not sure—remind me later	3
	4 (very important)

Answer	*Resilience Score*
No individual properties are equipped to either harvest rainwater or recharge groundwater.	1 (lowest resilience)
Few individual properties are equipped to either harvest rainwater or recharge groundwater.	2
Some individual properties are equipped to either harvest rainwater or recharge groundwater.	3
Most individual properties are equipped to either harvest rainwater or recharge groundwater.	4 (highest resilience)

#135: Are there redundant wastewater and stormwater systems in place for coping with extreme events, including collection systems and wastewater treatment systems?

Relevance	*Importance Weights*
Yes (relevant)	1 (not very important)
No (not relevant)	2
Not sure—remind me later	3
	4 (very important)

Answer	*Resilience Score*
No, redundant wastewater and stormwater systems are not in place.	1 (lowest resilience)
Yes, but these redundant wastewater and stormwater systems have only a small amount of the capacity necessary.	2
Yes, and these redundant wastewater and stormwater systems have some of the capacity necessary.	3
Yes, and these redundant wastewater and stormwater systems have all the capacity necessary.	4 (highest resilience)

#136: Does a water/wastewater agency response network provide technical resources/support to the urban area's water system during emergencies?

Relevance	*Importance Weights*
Yes (relevant)	1 (not very important)
No (not relevant)	2
Not sure—remind me later	3
	4 (very important)

Answer	*Resilience Score*
No	1 (lowest resilience)
Yes	3 (highest resilience)

#137: Have storm sewers and drains to storm sewers been inventoried, and are these inventories used in planning?

Relevance	*Importance Weights*
Yes (relevant)	1 (not very important)
No (not relevant)	2
Not sure—remind me later	3
	4 (very important)

Answer	*Resilience Score*
Storm sewers and drains to storm sewers are not inventoried and are not planned to be inventoried.	1 (lowest resilience)
Plans exist to inventory storm sewers and drains to storm sewers OR these inventories exist but are not used in planning.	2
Storm sewers and drains to storm sewers are being inventoried and these inventories are used or will be used in planning.	3
Storm sewers and drains to storm sewers have been inventoried and these inventories are used in planning.	4 (highest resilience)

#138: Is the availability of water goods and services at risk if other city goods and services (e.g., power, transportation, public health) are affected by extreme climatic events or gradual climatic changes?

Relevance	Importance Weights
Yes (relevant)	1 (not very important)
No (not relevant)	2
Not sure—remind me later	3
	4 (very important)

Answer	Resilience Score
Availability of water resources is at significant risk if other city services are affected by climatic events or changes.	1 (lowest resilience)
Availability of water resources is at moderate risk if other city services are affected by climatic events or changes.	2
Availability of water resources is at some risk if other city services are affected by climatic events or changes.	3
Availability of water resources is at minimal risk if other city services are affected by climatic events or changes.	4 (highest resilience)

#139: Has the water utility conducted a water audit to identify current losses (e.g., leaks, billing errors, inaccurate meters, unauthorized usage)?

Relevance	Importance Weights
Yes (relevant)	1 (not very important)
No (not relevant)	2
Not sure—remind me later	3
	4 (very important)

Answer	Resilience Score
No	1 (lowest resilience)
Yes	3 (highest resilience)

#140: To what extent have efforts been made to reduce water demand?

Relevance	*Importance Weights*
Yes (relevant)	1 (not very important)
No (not relevant)	2
Not sure—remind me later	3
	4 (very important)

Answer	*Resilience Score*
Few to no efforts have been made to reduce water demand.	1 (lowest resilience)
Fair efforts have been made to reduce water demand.	2
Moderate efforts have been made to reduce water demand.	3
Significant efforts have been made to reduce water demand.	4 (highest resilience)

#141: Are customers familiar with water conservation measures, and are they willing to implement these measures?

Relevance	*Importance Weights*
Yes (relevant)	1 (not very important)
No (not relevant)	2
Not sure—remind me later	3
	4 (very important)

Answer	*Resilience Score*
Customers are not familiar with OR are not willing to implement water conservation measures.	1 (lowest resilience)
Customers are marginally familiar with and somewhat or marginally willing to implement water conservation measures.	2
Customers are somewhat familiar with and willing to implement water conservation measures.	3
Customers are familiar with and willing to implement water conservation measures.	4 (highest resilience)

#163: Have water utility companies incorporated past experience or experience from other locations/utilities in developing plans for water shortages related to climate induced stresses?

Relevance	Importance Weights
Yes (relevant)	1 (not very important)
No (not relevant)	2
Not sure—remind me later	3
	4 (very important)

Answer	Resilience Score
No	1 (lowest resilience)
Yes	3 (highest resilience)

#164: Does the water department or utility for the city consider past experience in addressing anticipated increases in the frequency of sewer overflows?

Relevance	Importance Weights
Yes (relevant)	1 (not very important)
No (not relevant)	2
Not sure—remind me later	3
	4 (very important)

Answer	Resilience Score
No	1 (lowest resilience)
Yes	3 (highest resilience)

#166: Is backup power for wastewater collection and treatment provided?

Relevance	Importance Weights
Yes (relevant)	1 (not very important)
No (not relevant)	2
Not sure—remind me later	3
	4 (very important)

Answer	Resilience Score
No backup power is provided.	1 (lowest resilience)
Minimal backup power is provided.	2
Some backup power is provided.	3
Full backup power is provided.	4 (highest resilience)

APPENDIX J. QUANTITATIVE INDICATORS: TEMPLATE

A complete set of the quantitative indicators by sector developed for the tool.

J.1. Economy

The indicators below have been developed for the economy sector. Indicators that are related are grouped together such that a single indicator from that group was considered a **primary indicator** and the remaining were considered **secondary indicators.** Primary indicators and nongrouped indicators are presented in the first half of this handout, followed by the secondary indicators.

Each indicator has a **definition**. Each question is flagged with one or more of the following gradual change climate stressor and/or extreme event climate stressor (from the urban resilience framework developed for this project):

Stressors

Gradual Changes	*Extreme Events*
Wind speed	Magnitude/duration of heat waves
Temperature	Drought intensity/duration
Precipitation	Flood magnitude/frequency
Sea level rise	Hurricane intensity/frequency
	Storm surge/flooding

Where it was possible to identify a data set that would provide data for the indicator for Washington, DC **data sets** and associated **notes on available data** are included. Indicators are assigned a **proposed resilience score** on a scale of 1 (lowest resilience) to 4 (highest resilience).

For each indicator, please:

1. Discuss the **relevance** of the indicator to the economy sector. (If unsure, please select the *not sure—remind me later* option.) Indicators may be selected as *yes (relevant)* on the basis of the stressors previously selected as being most relevant to Washington, DC or based on any other criteria. Secondary indicators may be considered if the primary indicator is not adequately defined or does not have available data set(s).
2. When possible, **data sets** for Washington, DC are provided where data were available. In some cases, no data sets were identified. Please suggest data sets that may be better than the data sets identified or where data gaps exist.
3. For indicators selected as *yes (relevant)*, discuss an **importance weight**, where 1 = not very important and 4 = very important.

4. Review the **proposed resilience score** (if provided), which is on a scale of 1 (lowest resilience) to 4 (highest resilience), for the indicator. If you disagree with this score, please discuss **your score** and indicate the reason for your disagreement.

#709: Percentage of owned housing units that are affordable

Definition: This indicator measures (1) the percentage of owned housing units where selected monthly ownership costs (rent, mortgages, real estate taxes, insurance, utilities, fuel, fees) as a percentage of household income (SMOCAPI) exceeds 35% or (2) the percentage of rented housing units where gross rent as a percentage of household income (GRAPI) exceeds 35%.

Grouped with Indicators: N/A

Data Set(s):

Notes on Data Set(s):

Indicator Value:

Relevance:	*Importance Weight:*	*Proposed Resilience Score:*
Thresholds:	*Threshold-Based Score:*	*Your Score:*
0 to 30%	1 (lowest resilience)	1 (lowest resilience)
30 to 45%	2	2
45 to 60%	3	3
Greater than 60%	4 (highest resilience)	4 (highest resilience)

PRIMARY INDICATORS AND NONGROUPED INDICATORS

#717: Percentage access to health insurance of noninstitutionalized population

Definition: This indicator measures the percentage of noninstitutionalized residents with health insurance.

Grouped with Indicators: #725

Data Set(s):

Notes on Data Set(s):

Indicator Value:

Relevance:	*Importance Weight:*	*Proposed Resilience Score:*
Thresholds:	*Threshold-Based Score:*	*Your Score:*
Less than 85%	1 (lowest resilience)	1 (lowest resilience)
85 to 90%	2	2
90 to 95%	3	3
Greater than 95%	4 (highest resilience)	4 (highest resilience)

#711: Overall unemployment rate

Definition: Employment is a measure of economic viability and self-sufficiency. Employment opportunities spread across different industries create a more stable employment base. A diversification of industries also offers opportunities to a diverse labor market. This indicator measures the percentage of sectors in a city's economy that employ < 40% of the city's population. Sectors that employ 1% or less of the city's population are not considered, as they provide very minimal employment opportunities.

Grouped with Indicators: N/A

Data Set(s):

Notes on Data Set(s):

Indicator Value:

Relevance:	*Importance Weight:*	*Proposed Resilience Score:*
Thresholds:	*Threshold-Based Score:*	*Your Score:*
0 to less than 83%	1 (lowest resilience)	1 (lowest resilience)
83 to less than 91%	2	2
91 to less than 100%	3	3
100%	4 (highest resilience)	4 (highest resilience)

#722: Percentage change in homeless population

Definition: This indicator measures the percentage change in the homeless population.
Grouped with Indicators: N/A
Data Set(s):

Notes on Data Set(s):

Indicator Value:

Relevance:	Importance Weight:	Proposed Resilience Score:
Thresholds:	Threshold-Based Score:	Your Score:
Greater than 10%	1 (lowest resilience)	1 (lowest resilience)
1 to 10%	2	2
negative 10 to 0%	3	3
Less than negative 10%	4 (highest resilience)	4 (highest resilience)

#1375: Percentage of population living below the poverty line

Definition: This indicator measures the percentage of the population living below the poverty line.

Grouped with Indicators: N/A

Data Set(s):

Notes on Data Set(s):

Indicator Value:

Relevance:	*Importance Weight:*	*Proposed Resilience Score:*
Thresholds:	*Threshold-Based Score:*	*Your Score:*
Greater than 20%	1 (lowest resilience)	1 (lowest resilience)
16 to 20%	2	2
12 to 16%	3	3
Less than 12%	4 (highest resilience)	4 (highest resilience)

J.2. Energy

The indicators below have been developed for the energy sector. Indicators that are related are grouped together such that a single indicator from that group was considered a **primary indicator** and the remaining were considered **secondary indicators.** Primary indicators and nongrouped indicators are presented in the first half of this handout, followed by the secondary indicators.

Each indicator has a **definition**. Each question is flagged with one or more of the following gradual change climate stressor and/or extreme event climate stressor (from the urban resilience framework developed for this project):

Stressors

Gradual Changes
- Wind speed
- Temperature
- Precipitation
- Sea level rise

Extreme Events
- Magnitude/duration of heat waves
- Drought intensity/duration
- Flood magnitude/frequency
- Hurricane intensity/frequency
- Storm surge/flooding

Where it was possible to identify a data set that would provide data for the indicator for Washington, DC, **data sets** and associated **notes on available data** are included. Indicators are assigned a **proposed resilience score** on a scale of 1 (lowest resilience) to 4 (highest resilience).

For each indicator, please:

1. Discuss the **relevance** of the indicator to the energy sector. (If unsure, please select the *not sure—remind me later* option.) Indicators may be selected as *yes (relevant)* on the basis of the stressors previously selected as being most relevant to Washington, DC or based on any other criteria. Secondary indicators may be considered if the primary indicator is not adequately defined or does not have available data set(s).
2. When possible, **data sets** for Washington, DC are provided where data were available. In some cases, no data sets were identified. Please suggest data sets that may be better than the data sets identified or where data gaps exist.
3. For indicators selected as *yes (relevant)*, discuss an **importance weight**, where 1 = not very important and 4 = very important.
4. Review the **proposed resilience score** (if provided), which is on a scale of 1 (lowest resilience) to 4 (highest resilience), for the indicator. If you disagree with this score, please discuss **your score** and indicate the reason for your disagreement.

PRIMARY INDICATORS AND NONGROUPED INDICATORS

#949: Percentage energy consumed for electricity

Definition: The indicator measures electricity consumption per year in kWh as a percentage of total energy consumption.

Grouped with Indicators: #950, #951

Data Set(s):

Notes on Data Set(s):

Indicator Value:

Relevance:	*Importance Weight:*	*Proposed Resilience Score:*
Thresholds:	*Threshold-Based Score:* N/A	*Your Score:*
N/A	1 (lowest resilience)	1 (lowest resilience)
	2	2
	3	3
	4 (highest resilience)	4 (highest resilience)

#971: Energy source capacity per unit area

Definition: This indicator measures the total capacity of energy sources per unit area served (MW/sq mi).

Grouped with Indicators: #970

Data Set(s):

Notes on Data Set(s):

Indicator Value:

Relevance:	Importance Weight:	Proposed Resilience Score:
Thresholds:	Threshold-Based Score:	Your Score:
Less than 10 megawatts per square mile	1 (lowest resilience)	1 (lowest resilience)
10 to 50 megawatts per square mile	2	2
50 to 100 megawatts per square mile	3	3
Greater than 100 megawatts per square mile	4 (highest resilience)	4 (highest resilience)

#983: Average customer energy outage (hours) in recent major storm

Definition: This indicator measures the average customer energy outage hours divided by number of electricity customers for a storm event in June 2012.

Grouped with Indicators: #862

Data Set(s):

Notes on Data Set(s):

Indicator Value:

Alternate Data Set(s):

Notes on Alternate Data Set(s):

Alternate Indicator Value:

Relevance:	*Importance Weight:*	*Proposed Resilience Score:*
	1 (not very important)	
	2	
	3	
	4 (very important)	
Thresholds:	*Threshold-Based Score:*	*Your Score:*
Greater than 40 hours	1 (lowest resilience)	1 (lowest resilience)
20 to 40 hours	2	2
10 to 20 hours	3	3
Less than 10 hours	4 (highest resilience)	4 (highest resilience)

#898: Annual energy consumption per capita by main use category (commercial use)

Definition: The indicator measures the annual energy consumption (2010) per capita within the commercial use sector.

Grouped with Indicators: N/A

Data Set(s):

Notes on Data Set(s):

Indicator Value:

Alternate Data Set(s):

Notes on Alternate Data Set(s):

Alternate Indicator Value:

Relevance:	*Importance Weight:*	*Proposed Resilience Score:*
	1 (not very important)	
	2	
	3	
	4 (very important)	
Thresholds:	*Threshold-Based Score:*	*Your Score:*
Greater than 4.0 tons of oil equivalent	1 (lowest resilience)	1 (lowest resilience)
3.0 to 4.0 tons of oil equivalent	2	2
2.0 to 3.0 tons of oil equivalent	3	3
Less than or equal to 2.0 tons of oil equivalent	4 (highest resilience)	4 (highest resilience)

#967: Total energy source capacity per capita

Definition: This indicator measures the total capacity of all energy sources (MW) per capita.
Grouped with Indicators: N/A
Data Set(s):

Notes on Data Set(s):
Indicator Value:

Relevance:	*Importance Weight:*	*Proposed Resilience Score:*
Thresholds:	*Threshold-Based Score:*	*Your Score:*
Less than 1.0 megawatt per capita	1 (lowest resilience)	1 (lowest resilience)
1.0 to 2.0 megawatts per capita	2	2
2.0 to 5.0 megawatts per capita	3	3
Greater than 5.0 megawatts per capita	4 (highest resilience)	4 (highest resilience)

SECONDARY INDICATORS

#950: Percentage of electricity generation from noncarbon sources

Definition: This indicator measures the percentage of total electricity generation from noncarbon energy sources in a city.

Grouped with Indicators: #949, #951

Data Set(s):

Notes on Data Set(s):

Indicator Value:

Relevance:	Importance Weight:	Proposed Resilience Score:
Thresholds:	Threshold-Based Score:	Your Score:
Less than 25%	1 (lowest resilience)	1 (lowest resilience)
25 to 50%	2	2
50 to 75%	3	3
Greater than 75%	4 (highest resilience)	4 (highest resilience)

#951: Percentage of total energy use from renewable sources

Definition: This indicator measures the percentage of total energy use from renewable sources.

Grouped with Indicators: #949, #950

Data Set(s):

Notes on Data Set(s):

Indicator Value:

Alternate Data Set(s):

Notes on Alternate Data Set(s):

Alternate Indicator Value:

Relevance:	*Importance Weight:*	*Proposed Resilience Score:*
	3	
	1 (not very important)	
	2	
	3	
	4 (very important)	
Thresholds:	*Threshold-Based Score:*	*Your Score:*
Less than 20%	1 (lowest resilience)	1 (lowest resilience)
20 to 40%	2	2
40 to 60%	3	3
Greater than 60%	4 (highest resilience)	4 (highest resilience)

#970: Average capacity of a decentralized energy source

Definition: This indicator measures the average capacity of a decentralized energy source (m^3/acre). Decentralized energy sources are those that can be used as a supplementary source to the existing centralized energy system. They are typically located closer to the site of actual energy consumption than centralized sources.

Grouped with Indicators: #971

Data Set(s):

Notes on Data Set(s):

Indicator Value:

Relevance:	*Importance Weight:*	*Proposed Resilience Score:*
Not Sure		
Yes (relevant)	1 (not very important)	
No (not relevant)	2	
	3	
	4 (very important)	
Thresholds:	*Threshold-Based Score:* N/A	*Your Score:*
Less than 5,000 megawatts per square mile	1 (lowest resilience)	1 (lowest resilience)
5,000 to 10,000 megawatts per square mile	2	2
10,000 to 15,000 megawatts per square mile	3	3
Greater than 15,000 megawatts per square mile	4 (highest resilience)	4 (highest resilience)

#924: Energy intensity by use

Definition: This indicator measures energy intensity in manufacturing, transportation, agriculture, commercial and public services, and the residential sector.

Grouped with Indicators:

Data Set(s):

Notes on Data Set(s):

Indicator Value:

Relevance:	*Importance Weight:*	*Proposed Resilience Score:*
Thresholds:	*Threshold-Based Score:*	*Your Score:*
Greater than 3,000 Btu per dollar	1 (lowest resilience)	1 (lowest resilience)
2,000 to 3,000 Btu per dollar	2	2
1,500 to 2,000 Btu per dollar	3	3
Less than 1,500 Btu per dollar	4 (highest resilience)	4 (highest resilience)

J.3. Land Use/Land Cover

The indicators below have been developed for the land use/land cover sector. Indicators that are related are grouped together such that a single indicator from that group was considered a **primary indicator** and the remaining were considered **secondary indicators.** Primary indicators and nongrouped indicators are presented in the first half of this handout, followed by the secondary indicators.

Each indicator has a **definition**. Each question is flagged with one or more of the following gradual change climate stressor and/or extreme event climate stressor (from the urban resilience framework developed for this project):

Stressors

Gradual Changes
- Wind speed
- Temperature
- Precipitation
- Sea level rise

Extreme Events
- Magnitude/duration of heat waves
- Drought intensity/duration
- Flood magnitude/frequency
- Hurricane intensity/frequency
- Storm surge/flooding

Where it was possible to identify a data set that would provide data for the indicator for Washington, DC, **data sets** and associated **notes on available data** are included. Indicators are assigned a **proposed resilience score** on a scale of 1 (lowest resilience) to 4 (highest resilience).

For each indicator, please:

1. Discuss the **relevance** of the indicator to the land use/land cover sector. (If unsure, please select the *not sure—remind me later* option.) Indicators may be selected as *yes (relevant)* on the basis of the stressors previously selected as being most relevant to Washington, DC or based on any other criteria. Secondary indicators may be considered if the primary indicator is not adequately defined or does not have available data set(s).
2. When possible, **data sets** for Washington, DC are provided where data were available. In some cases, no data sets were identified. Please suggest data sets that may be better than the data sets identified or where data gaps exist.
3. For indicators selected as *yes (relevant)*, discuss an **importance weight**, where 1 = not very important and 4 = very important.
4. Review the **proposed resilience score** (if provided), which is on a scale of 1 (lowest resilience) to 4 (highest resilience), for the indicator. If you disagree with this score, please discuss **your score** and indicate the reason for your disagreement.

PRIMARY INDICATORS AND NONGROUPED INDICATORS

#437: Percentage change in streamflow divided by percentage change in precipitation

Definition: The proportional change in streamflow (Q) divided by the proportional change in precipitation (P) for 1,291 gauged watersheds across the continental U.S. from 1931 to 1988.

Grouped with Indicators: #1369

Data Set(s):

Notes on Data Set(s):

Indicator Value:

Relevance:	*Importance Weight:*	*Proposed Resilience Score:*
Thresholds:	*Threshold-Based Score:*	*Your Score:*
Greater than 3.0 (unitless ratio)	1 (lowest resilience)	1 (lowest resilience)
2.0 to 3.0 (unitless ratio)	2	2
1.0 to 2.0 (unitless ratio)	3	3
Less than 1.0 (unitless ratio)	4 (highest resilience)	4 (highest resilience)

#825: Percentage change in impervious cover

Definition: This indicator reflects the change in the percentage of the metropolitan area that is impervious surface (roads, buildings, sidewalks, parking lots, etc.).

Grouped with Indicators: #303, #308

Data Set(s):

Notes on Data Set(s):

Indicator Value:

Alternate Data Set(s):

Notes on Alternate Data Set(s):

Alternate Indicator Value:

Relevance:	*Importance Weight:*	*Proposed Resilience Score:*
	1 (not very important)	
	2	
	3	
	4 (very important)	
Thresholds:	*Threshold-Based Score:*	*Your Score:*
Greater than 1%	1 (lowest resilience)	1 (lowest resilience)
0 to 1%	2	2
Negative 1 to 0%	3	3
Less than negative 1%	4 (highest resilience)	4 (highest resilience)

#1436: Percentage of city area in 100-year floodplain

Definition: This indicator reflects the percentage of the metropolitan area that lies within the 100-year floodplain.

Grouped with Indicators: #1437, #1438, #1439

Data Set(s):

Notes on Data Set(s):

Indicator Value:

Relevance:	*Importance Weight:*	*Proposed Resilience Score:*
Thresholds:	*Threshold-Based Score:*	*Your Score:*
Greater than 20%	1 (lowest resilience)	1 (lowest resilience)
5 to 20%	2	2
1 to 5%	3	3
Less than 1%	4 (highest resilience)	4 (highest resilience)

#51: Coastal Vulnerability Index rank

Definition: This indicator reflects the Coastal Vulnerability Index rank. The ranks are as follows: 1 = none, 2 = low, 3 = moderate, 4 = high, 5 = very high. The index allows six physical variables to be related in a quantifiable manner that expresses the relative vulnerability of the coast to physical changes due to sea level rise. The six variables are: a = geomorphology, b = coastal slope (%), c = relative sea level change (mm/year), d = shoreline erosion/accretion (m/year), e = mean tide average (m), and f = mean wave height (m).

Grouped with Indicators: N/A

Data Set(s):

Notes on Data Set(s):

Indicator Value:

Relevance:	*Importance Weight:*	*Proposed Resilience Score:*
Thresholds:	*Threshold-Based Score:*	*Your Score:*
5 (very high vulnerability)	1 (lowest resilience)	1 (lowest resilience)
4 (high vulnerability)	2	2
3 (moderate vulnerability)	3	3
Less than or equal to 2 (low or no vulnerability)	4 (highest resilience)	4 (highest resilience)

#194: Percentage of natural area that is in small natural patches

Definition: This indicator measures the percentage of the total natural area in a city that is in patches of less than 10 acres. Smaller patches of natural habitat generally provide lower-quality habitat for plants and animals and provide less solitude and fewer recreational opportunities for people. About half of all natural lands in urban and suburban areas are in patches smaller than 10 acres.

Grouped with Indicators: N/A

Data Set(s):

Notes on Data Set(s):

Indicator Value:

Relevance:	Importance Weight:	Proposed Resilience Score:
Yes (relevant)	1 (not very important)	
No (not relevant)	2	
Not sure—remind me later	3	
	4 (very important)	

Thresholds:	Threshold-Based Score:	Your Score:
Greater than 80%	1 (lowest resilience)	1 (lowest resilience)
60 to 80%	2	2
40 to 60%	3	3
Less than 40%	4 (highest resilience)	4 (highest resilience)

#254: Ratio of perimeter to area of natural patches

Definition: This indicator is calculated as the average ratio of the perimeter to area.
Grouped with Indicators: N/A
Data Set(s):

Notes on Data Set(s):

Indicator Value:

Relevance:	*Importance Weight:*	*Proposed Resilience Score:*
Yes (relevant)	1 (not very important)	
No (not relevant)	2	
Not sure—remind me later	3	
	4 (very important)	
Thresholds:	*Threshold-Based Score:*	*Your Score:*
Greater than 0.025 (unitless ratio)	1 (lowest resilience)	1 (lowest resilience)
0.015 to 0.025 (unitless ratio)	2	2
0.005 to 0.015 (unitless ratio)	3	3
Less than 0.005 (unitless ratio)	4 (highest resilience)	4 (highest resilience)

#1440: Palmer Drought Severity Index

Definition:
(1) Calculate potential evapotranspiration (PET) for selected time periods using temperature data and the Thornthwaite equation.
(2) Find the precipitation deficit (precipitation minus PET) for the selected time period, where more negative values indicate greatest precipitation deficit.
(3) Using a moving window sum, find the 1-, 3-, 6-, or 12-month period that had the greatest total precipitation deficit.

Grouped with Indicators: N/A

Data Set(s):

Notes on Data Set(s):

Indicator Value:

Relevance:	Importance Weight:	Proposed Resilience Score:
Thresholds:	Threshold-Based Score:	Your Score:
Less than or equal to negative 4.0 (extreme drought)	1 (lowest resilience)	1 (lowest resilience)
Negative 3.99 to negative 3.0 (severe drought)	2	2
Negative 2.99 to negative 2.0 (moderate drought)	3	3
Greater than or equalt to negative 1.99 (mild or no drought)	4 (highest resilience)	4 (highest resilience)

SECONDARY INDICATORS

#308: Percentage of land that is urban/suburban

Definition: This indicator presents the extent/acreage of urban and suburban areas as a percentage of the total U.S. land area, for the most recent 50-year period and compared to presettlement estimates. It also reports on a key component of freshwater ecosystems (freshwater wetlands) and will report on the area of brackish water, a key component of coastal and ocean ecosystems when data become available.

Grouped with Indicators: #303, #825

Relevance:	*Importance Weights:*
Yes (relevant)	1 (not very important)
No (not relevant)	2
Not sure—remind me later	3
	4 (very important)

Data Set(s):

Notes on Data Set(s):

Indicator Value:

Proposed Resilience Score:	*Your Score:*
	1 (lowest resilience)
	2
	3
	4 (highest resilience)

#1369: Annual CV of unregulated streamflow

Definition: The coefficient of variation (CV) of unregulated streamflow is an indicator of annual streamflow variability. It is computed as the ratio of the standard deviation of unregulated annual streamflow (oQs) to the unregulated mean annual streamflow (QS)' (Hurd et al., 1999).

Grouped with Indicators: #437

Data Set(s):

Notes on Data Set(s):

Indicator Value:

Relevance: *Importance Weight:* *Proposed Resilience Score:*

Thresholds:	Threshold-Based Score:	Your Score:
Greater than 0.60 (unitless ratio)	1 (lowest resilience)	1 (lowest resilience)
0.40 to 0.60 (unitless ratio)	2	2
0.20 to 0.40 (unitless ratio)	3	3
Less than 0.20 (unitless ratio)	4 (highest resilience)	4 (highest resilience)

#1437: Percentage of city area in 500-year floodplain

Definition: This indicator reflects the percentage of the metropolitan area that lies within the 500-year floodplain.

Grouped with Indicators: #1436, #1438, #1439

Data Set(s):

Notes on Data Set(s):

Indicator Value:

Relevance:	*Importance Weight:*	*Proposed Resilience Score:*
Thresholds:	*Threshold-Based Score:*	*Your Score:*
Greater than 30%	1 (lowest resilience)	1 (lowest resilience)
10 to 30%	2	2
2 to 10%	3	3
Less than 2%	4 (highest resilience)	4 (highest resilience)

#1438: Percentage of city population in 100-year floodplain

Definition: This indicator reflects the percentage of the city population living within the 100-year floodplain.

Grouped with Indicators: #1436, #1437, #1439

Data Set(s):

Notes on Data Set(s):

Indicator Value:

Relevance:	Importance Weight:	Proposed Resilience Score:
Thresholds:	Threshold-Based Score:	Your Score:
Greater than 20%	1 (lowest resilience)	1 (lowest resilience)
5 to 20%	2	2
1 to 5%	3	3
Less than 1%	4 (highest resilience)	4 (highest resilience)

#1439: Percentage of city population in 500-year floodplain

Definition: This indicator reflects the percentage of the city population living within the 500-year floodplain.

Grouped with Indicators: #1436, #1437, #1438

Data Set(s):

Notes on Data Set(s):

Indicator Value:

Relevance:	Importance Weight:	Proposed Resilience Score:
Thresholds:	Threshold-Based Score:	Your Score:
Greater than 30%	1 (lowest resilience)	1 (lowest resilience)
10 to 30%	2	2
2 to 10%	3	3
Less than 2%	4 (highest resilience)	4 (highest resilience)

J.4. Natural Environment

The indicators below have been developed for the natural environment sector. Indicators that are related are grouped together such that a single indicator from that group was considered a **primary indicator** and the remaining were considered **secondary indicators.** Primary indicators and nongrouped indicators are presented in the first half of this handout, followed by the secondary indicators.

Each indicator has a **definition**. Each question is flagged with one or more of the following gradual change climate stressor and/or extreme event climate stressor (from the urban resilience framework developed for this project):

Stressors

Gradual Changes
- Wind speed
- Temperature
- Precipitation
- Sea level rise

Extreme Events
- Magnitude/duration of heat waves
- Drought intensity/duration
- Flood magnitude/frequency
- Hurricane intensity/frequency
- Storm surge/flooding

Where it was possible to identify a data set that would provide data for the indicator for Washington, DC, **data sets** and associated **notes on available data** are included. Indicators are assigned a **proposed resilience score** on a scale of 1 (lowest resilience) to 4 (highest resilience).

For each indicator, please:

1. Discuss the **relevance** of the indicator to the natural environment sector. (If unsure, please select the *not sure—remind me later* option.) Indicators may be selected as *yes (relevant)* on the basis of the stressors previously selected as being most relevant to Washington, DC or based on any other criteria. Secondary indicators may be considered if the primary indicator is not adequately defined or does not have available data set(s).
2. When possible, **data sets** for Washington, DC are provided where data were available. In some cases, no data sets were identified. Please suggest data sets that may be better than the data sets identified or where data gaps exist.
3. For indicators selected as *yes (relevant)*, discuss an **importance weight**, where 1 = not very important and 4 = very important.
4. Review the **proposed resilience score** (if provided), which is on a scale of 1 (lowest resilience) to 4 (highest resilience), for the indicator. If you disagree with this score, please discuss **your score** and indicate the reason for your disagreement.

PRIMARY INDICATORS AND NONGROUPED INDICATORS

#682: Percentage change in bird population

Definition: This indicator reflects the number of species with "substantial" increases or decreases in the number of observations (not a change in the number of species) divided by the total number of bird species.

Grouped with Indicators: #680, #681

Data Set(s):

Notes on Data Set(s):

Indicator Value:

Relevance:	*Importance Weight:*	*Proposed Resilience Score:*
Thresholds:	*Threshold-Based Score:*	*Your Score:*
Less than negative 66%	1 (lowest resilience)	1 (lowest resilience)
Negative 66 to 0%	2	2
0 to 66%	3	3
Greater than 66%	4 (highest resilience)	4 (highest resilience)

#17: Altered wetlands (percentage of wetlands lost)

Definition: This indicator reflects the percentage of wetland areas that have been excavated, impounded, diked, partially drained, or farmed.

Grouped with Indicators: N/A

Data Set(s):

Notes on Data Set(s):

Indicator Value:

Relevance:	*Importance Weight:*	*Proposed Resilience Score:*
Thresholds:	*Threshold-Based Score:*	*Your Score:*
Greater than 60%	1 (lowest resilience)	1 (lowest resilience)
40 to 60%	2	2
20 to 40%	3	3
Less than 20%	4 (highest resilience)	4 (highest resilience)

#66: Percentage change in disruptive species

Definition: This indicator reflects the percentage change in disruptive species found in metropolitan areas. Disruptive species are those that have negative effects on natural areas and native species or cause damage to people and property.

Grouped with Indicators: N/A

Data Set(s):

Notes on Data Set(s):

Indicator Value:

Relevance:	*Importance Weight:*	*Proposed Resilience Score:*
Yes (relevant)		
No (not relevant)		
Not sure—remind me later		
Thresholds:	*Threshold-Based Score:*	*Your Score:*
Greater than 100%	1 (lowest resilience)	1 (lowest resilience)
50 to 100%	2	2
10 to 50%	3	3
Less than 10%	4 (highest resilience)	4 (highest resilience)

#273: Percentage of total wildlife species of greatest conservation need

Definition: This indicator reflects the percentage of total wildlife species that are listed as having the "greatest conservation need."

Grouped with Indicators: N/A

Data Set(s):

Notes on Data Set(s):

Indicator Value:

Relevance:	Importance Weight:	Proposed Resilience Score:
Thresholds:	*Threshold-Based Score:*	*Your Score:*
Greater than 20%	1 (lowest resilience)	1 (lowest resilience)
5 to 20%	2	2
1 to 5%	3	3
Less than 1%	4 (highest resilience)	4 (highest resilience)

#284: Physical Habitat Index (PHI)

Definition: PHI includes eight characteristics (riffle quality, stream bank stability, quantity of woody debris, instream habitat for fish, suitability of streambed surface materials for macroinvertebrates, shading, distance to nearest road, and embeddedness of substrates). Scores range from 0–100 (81–100 = minimally degraded, 66–80 = partially degraded, 51–65 = degraded, 0–50 = severely degraded).

Grouped with Indicators: N/A

Data Set(s):

Indicator Value:

Relevance:	*Importance Weight:*	*Proposed Resilience Score:*
Thresholds:	*Threshold-Based Score:*	*Your Score:*
0 to 50 (severely degraded)	1 (lowest resilience)	1 (lowest resilience)
51 to 65 (degraded)	2	2
66 to 80 (partially degraded)	3	3
81 to 100 (minimally degraded)	4 (highest resilience)	4 (highest resilience)

#326: Wetland species at risk (number of species)

Definition: Number of wetland and freshwater species at risk (rare, threatened, or endangered).
Grouped with Indicators: N/A
Data Set(s):

Notes on Data Set(s):

Indicator Value:

Relevance:	*Importance Weight:*	*Proposed Resilience Score:*
Thresholds:	*Threshold-Based Score:*	*Your Score:*
Greater than 160 species at risk	1 (lowest resilience)	1 (lowest resilience)
100 to 160 species at risk	2	2
50 to less than 100 species at risk	3	3
Less than 50 species at risk	4 (highest resilience)	4 (highest resilience)

#460: Macroinvertebrate Index of Biotic Condition

Definition: The Benthic Index of Biotic Integrity (BIBI) score is the average of the score of 10 individual metrics, including Total Taxa Richness, Ephemeroptera Taxa Richness, Plecoptera Taxa Richness, Trichoptera Taxa Richness, Intolerant Taxa Richness, Clinger Taxa Richness and Percentage, Long-Lived Taxa Richness, Percentage Tolerant, Percentage Predator, and Percentage Dominance.

Grouped with Indicators: N/A

Data Set(s):

Notes on data sets(s):

Indicator Value:

Relevance:	*Importance Weight:*	*Proposed Resilience Score:*
Thresholds:	*Threshold-Based Score:*	*Your Score:*
0 to 45 (poor or very poor biotic condition)	1 (lowest resilience)	1 (lowest resilience)
46 to 55 (fair biotic condition)	2	2
56 to 75 (good biotic condition)	3	3
Greater than 75 (very good biotic condition)	4 (highest resilience)	4 (highest resilience)

#465: Change in plant species diversity from pre-European settlement

Definition: Change in the plant species diversity from pre-European settlement (baseline) to present, within a given city/area.

Grouped with Indicators: N/A

Data Set(s):

Notes on Data Set(s):

Indicator Value:

Relevance:	Importance Weight:	Proposed Resilience Score:
Thresholds:	Threshold-Based Score:	Your Score:
Less than 0.2 Shannon Diversity Index	1 (lowest resilience)	1 (lowest resilience)
0.2 to 0.4 Shannon Diversity Index	2	2
0.4 to 0.6 Shannon Diversity Index	3	3
Greater than 0.60 Shannon Diversity Index	4 (highest resilience)	4 (highest resilience)

SECONDARY INDICATORS

#680: Ecological connectivity (percentage of area classified as hub or corridor)

Definition: This indicator reflects the percentage of the metropolitan area identified as a "hub" or "corridor." Hubs are large areas of important natural ecosystems such as the Okefenokee National Wildlife Refuge in Georgia and the Osceola National Forest in Florida. Corridors (i.e., "connections") are links to support the functionality of the hubs (e.g., the Pinhook Swamp which connects the Okefenokee and Osceola hubs).

Grouped with Indicators: #681, #682

Data Set(s):

Notes on Data Set(s):

Indicator Value:

Relevance:	*Importance Weight:*	*Proposed Resilience Score:*
Thresholds:	*Threshold-Based Score:* N/A	*Your Score:* 2
Less than 10%	1 (lowest resilience)	1 (lowest resilience)
10 to 25%	2	2
25 to 50%	3	3
Greater than 50%	4 (highest resilience)	4 (highest resilience)

#681: Relative ecological condition of undeveloped land

Definition: This indicator characterizes the ecological condition of undeveloped land based on three indices derived from criteria representing diversity, self-sustainability, the rarity of certain types of land cover, species, and higher taxa (White and Maurice, 2004). In this context, "undeveloped land" refers to all land use not classified as urban, industrial, residential, or agricultural.

Grouped with Indicators: #680, #682

Data Set(s):

Notes on Data Set(s):

Indicator Value:

Relevance:	*Importance Weight:*	*Proposed Resilience Score:*
Thresholds:	*Threshold-Based Score:* N/A	*Your Score:*
Less than 120 White and Maurice Index score	1 (lowest resilience)	1 (lowest resilience)
120 to 180 White and Maurice Index score	2	2
180 to 230 White and Maurice Index score	3	3
Greater than 230 White and Maurice Index score	4 (highest resilience)	4 (highest resilience)

J.5. People

The indicators below have been developed for the people sector. Indicators that are related are grouped together such that a single indicator from that group was considered a **primary indicator** and the remaining were considered **secondary indicators**. Primary indicators and nongrouped indicators are presented in the first half of this handout, followed by the secondary indicators.

Each indicator has a **definition**. Each question is flagged with one or more of the following gradual change climate stressor and/or extreme event climate stressor (from the urban resilience framework developed for this project):

Stressors

Gradual Changes	*Extreme Events*
Wind speed	Magnitude/duration of heat waves
Temperature	Drought intensity/duration
Precipitation	Flood magnitude/frequency
Sea level rise	Hurricane intensity/frequency
	Storm surge/flooding

Where it was possible to identify a data set that would provide data for the indicator for Washington, DC, **data sets** and associated **notes on available data** are included. Indicators are assigned a **proposed resilience score** on a scale of 1 (lowest resilience) to 4 (highest resilience).

For each indicator, please:

1. Discuss the **relevance** of the indicator to the people sector. (If unsure, please select the *not sure—remind me later* option.) Indicators may be selected as *yes (relevant)* on the basis of the stressors previously selected as being most relevant to Washington, DC or based on any other criteria. Secondary indicators may be considered if the primary indicator is not adequately defined or does not have available data set(s).
2. When possible, **data sets** for Washington, DC are provided where data were available. In some cases, no data sets were identified. Please suggest data sets that may be better than the data sets identified or where data gaps exist.
3. For indicators selected as *yes (relevant)*, discuss an **importance weight**, where 1 = not very important and 4 = very important.
4. Review the **proposed resilience score** (if provided), which is on a scale of 1 (lowest resilience) to 4 (highest resilience), for the indicator. If you disagree with this score, please discuss **your score** and indicate the reason for your disagreement.

PRIMARY INDICATORS AND NONGROUPED INDICATORS

#675: Asthma Prevalence (Percentage of population affected by asthma)

Definition: This indicator presents asthma prevalence for U.S. children (age 0-17) and adults (age 18 and older). It is calculated as the percentage of population reporting asthma. Asthma attack prevalence is based on the number of adults/children who reported an asthma episode or attack in the past 12 months.

Grouped with Indicators: N/A

Data Set(s):

Notes on Data Set(s):

Indicator Value:

Relevance:	Importance Weight:	Proposed Resilience Score:
Thresholds:	Threshold-Based Score:	Your Score:
Greater than 12%	1 (lowest resilience)	1 (lowest resilience)
9 to 12%	2	2
6 to 9%	3	3
Less than 6%	4 (highest resilience)	4 (highest resilience)

#676: Percentage of population affected by notifiable diseases

Definition: This indicator reflects percentage occurrence of notifiable diseases as reported by health departments to the National Notifiable Diseases Surveillance System (NNDSS). A notifiable disease is one for which regular, frequent, and timely information regarding individual cases is considered necessary for the prevention and control of the disease (CDC, 2005b). The "notifiable diseases" included are chlamydia, coccidioidomycosis, cryptosporidiosis, Dengue virus, *Escherichia coli*, ehrlichiosis, Giardiasis, gonorrhea, *Haemophilus influenzae*, hepatitus A, hepatitus B, hepatitus C, legionellosis, Lyme disease, malaria, meningococcal disease, mumps, pertussis (whooping cough), rabies, Salmonellosis, shigellosis, spotted fever rickettsiosis/Rocky Mountain spotted fever, *Streptococcus pneumoniae*, syphilis, tuberculosis, varicella (chicken pox), and West Nile/meningitis/encephalitis.

Grouped with Indicators: #322, #1171

Data Set(s):

Notes on Data Set(s):

Indicator Value:

Relevance:	*Importance Weight:*	*Proposed Resilience Score:*
Thresholds:	*Threshold-Based Score:*	*Your Score:*
Greater than 3 to 4%	1 (lowest resilience)	1 (lowest resilience)
2 to 3%	2	2
1 to 2%	3	3
Less than 1%	4 (highest resilience)	4 (highest resilience)

#690: Emergency medical service response times

Definition: This indicator measures average annual response times (in minutes) for emergency medical service calls.

Grouped with Indicators: #757, #784, #798

Data Set(s):

Notes on Data Set(s):

Indicator Value:

Relevance:	Importance Weight:	Proposed Resilience Score:
Thresholds:	*Threshold-Based Score:*	*Your Score:*
Greater than 12 minutes	1 (lowest resilience)	1 (lowest resilience)
10 to 12 minutes	2	2
8 to 10 minutes	3	3
Less than 8 minutes	4 (highest resilience)	4 (highest resilience)

#1387: Percentage of population vulnerable due to age

Definition: This indicator reflects percentage of population above 65 or under 5 years old.
Grouped with Indicators: #393, #728. #1157, #1170

Data Set(s):

Notes on Data Set(s):

Indicator Value:

Relevance:	Importance Weight:	Proposed Resilience Score:
Thresholds:	Threshold-Based Score:	Your Score:
Greater than 20%	1 (lowest resilience)	1 (lowest resilience)
15 to 20%	2	2
10 to 15%	3	3
Less than 10%	4 (highest resilience)	4 (highest resilience)

#209: Percentage of population living within the 500-year floodplain

Definition: This indicator reflects percentage of population living within the 500-year floodplain.

Grouped with Indicators: N/A

Data Set(s):

Notes on Data Set(s):

Indicator Value:

Relevance:	*Importance Weight:*	*Proposed Resilience Score:*
Thresholds:	*Threshold-Based Score:*	*Your Score:*
Greater than 30%	1 (lowest resilience)	1 (lowest resilience)
10 to 30%	2	2
2 to 10%	3	3
Less than 2%	4 (highest resilience)	4 (highest resilience)

#725: Number of physicians per capita

Definition: This indicator reflects the total number of M.D. and D.O. physicians per capita.

Grouped with Indicators: #717

Data Set(s):

Notes on Data Set(s):

Indicator Value:

Relevance:	*Importance Weight:*	*Proposed Resilience Score:*
Thresholds:	*Threshold-Based Score:*	*Your Score:*
Less than 0.02 physicians per capita	1 (lowest resilience)	1 (lowest resilience)
0.02 to 0.03 physicians per capita	2	2
0.03 to 0.04 physicians per capita	3	3
Greater than 0.04 physicians per capita	4 (highest resilience)	4 (highest resilience)

#1376: Percentage of population that is disabled

Definition: This indicator reflects the percentage of the noninstitutionalized population that is disabled. Disabled individuals are those who have one or more of the following: hearing difficulty (deaf or having serious difficulty hearing), vision difficulty (blind or having serious difficulty seeing, even when wearing glasses), cognitive difficulty (having difficulty remembering, concentrating, or making decisions because of a physical, mental, or emotional problem), ambulatory difficulty (serious difficulty walking or climbing stairs), self-care difficulty (difficulty bathing or dressing), and independent living difficulty (difficulty doing errands because of a physical, mental, or emotional problem).

Grouped with Indicators: N/A

Data Set(s):

Notes on Data Set(s):

Indicator Value:

Relevance:	*Importance Weight:*	*Proposed Resilience Score:*
Thresholds:	*Threshold-Based Score:*	*Your Score:*
Greater than 20%	1 (lowest resilience)	1 (lowest resilience)
15 to 20%	2	2
10 to 15%	3	3
Less than 10%	4 (highest resilience)	4 (highest resilience)

#1390: Percentage of population that is living alone

Definition: This indicator reflects the percentage of population that is 65 years or older and living alone.

Grouped with Indicators: N/A

Data Set(s):

Notes on Data Set(s):

Indicator Value:

Relevance:	Importance Weight:	Proposed Resilience Score:
Thresholds:	Threshold-Based Score:	Your Score:
Greater than 30%	1 (lowest resilience)	1 (lowest resilience)
20 to 30%	2	2
10 to 20%	3	3
Less than 10%	4 (highest resilience)	4 (highest resilience)

#1443: Deaths from extreme weather events

Definition: This indicator measures the number of deaths in the last 5 years due to extreme events (cold, flood, heat, lightning, tornado, tropical cyclone, wind, and winter storms).

Grouped with Indicators: N/A

Data Set(s):

Notes on Data Set(s):

Indicator Value:

Relevance:	*Importance Weight:*	*Proposed Resilience Score:*
Thresholds:	*Threshold-Based Score:*	*Your Score:*
Greater than 150 deaths	1 (lowest resilience)	1 (lowest resilience)
100 to 150 deaths	2	2
50 to 100 deaths	3	3
Less than 50 deaths	4 (highest resilience)	4 (highest resilience)

SECONDARY INDICATORS

#322: Percentage of population affected by waterborne diseases

Definition: This indicator reports the percentage of population affected by waterborne diseases.

Grouped with Indicators: #676, #1171

Data Set(s):

Notes on Data Set(s):

Indicator Value:

Relevance:	*Importance Weight:*	*Proposed Resilience Score:*
Thresholds:	*Threshold-Based Score:*	*Your Score:*
Greater than 2%	1 (lowest resilience)	1 (lowest resilience)
1 to 2%	2	2
0 to 1%	3	3
0%	4 (highest resilience)	4 (highest resilience)

#393: Percentage of vulnerable population that is homeless

Definition: This indicator reflects the percentage of population 65 and older and under 5 years that is homeless.

Grouped with Indicators: #728, #1157, #1170, #1387

Data Set(s):

Notes on Data Set(s):

Indicator Value:

Relevance:	Importance Weight:	Proposed Resilience Score:
Thresholds:	Threshold-Based Score:	Your Score:
Greater than 30%	1 (lowest resilience)	1 (lowest resilience)
20 to 30%	2	2
10 to 20%	3	3
Less than 10%	4 (highest resilience)	4 (highest resilience)

#728: Adult care (homes per capita)

Definition: The number of adult day care homes and assisted living homes per capita of population over 65 years.

Grouped with Indicators: #393, #1157, #1170, #1387

Data Set(s):

Notes on Data Set(s):

Indicator Value:

Alternate Data Set(s):

Notes on Alternate Data Set(s):

Alternate Indicator Value:

Relevance:	*Importance Weight:*	*Proposed Resilience Score:*
	1 (not very important)	
	2	
	3	
	4 (very important)	
Thresholds:	*Threshold-Based Score:*	*Your Score:*
Less than 0.00010 adult homes per capita of elderly population	1 (lowest resilience)	1 (lowest resilience)
0.00010 to 0.00020 adult homes per capita of elderly population	2	2
0.00020 to 0.00040 adult homes per capita of elderly population	3	3
Greater than 0.00040 adult homes per capita of elderly population	4 (highest resilience)	4 (highest resilience)

#757: Average police response time

Definition: This indicator reflects the average response time for police to respond to emergency situations.

Grouped with Indicators: #690, #784, #798

Data Set(s):

Notes on Data Set(s):

Indicator Value:

Relevance:	Importance Weight:	Proposed Resilience Score:
	1 (not very important)	
	2	
	3	
	4 (very important)	
Thresholds:	*Threshold-Based Score:*	*Your Score:*
Greater than 12 minutes	1 (lowest resilience)	1 (lowest resilience)
10 to 12 minutes	2	2
8 to 10 minutes	3	3
Less than 8 minutes	4 (highest resilience)	4 (highest resilience)

#784: Number of sworn police officers per capita

Definition: This indicator is calculated by dividing the number of sworn police officers by the total population. We multiply the result by 1,000. According to the FBI, sworn officers meet the following criteria: "they work in an official capacity, they have full arrest powers, they wear a badge (ordinarily), they carry a firearm (ordinarily), and they are paid from governmental funds set aside specifically for payment of sworn law enforcement representatives." In counties with relatively few people, a small change in the number of officers may have a significant effect on rates from year to year.

Grouped with Indicators: #690, #757, #798

Data Set(s):

Notes on Data Set(s):

Indicator Value:

Relevance:	Importance Weight:	Proposed Resilience Score:
Thresholds:	Threshold-Based Score:	Your Score:
Less than 0.10 police officers per capita	1 (lowest resilience)	1 (lowest resilience)
0.10 to 0.20 police officers per capita	2	2
0.20 to 0.50 police officers per capita	3	3
Greater than 0.50 police officers per capita	4 (highest resilience)	4 (highest resilience)

#798: Percentage of fire response times less than 6.5 minutes

Definition: This indicator reflects the percentage of fire response times less than 6.5 minutes (from city stations to city locations).

Grouped with Indicators: #690, #757, #784

Data Set(s):

Notes on Data Set(s):

Indicator Value:

Relevance:	Importance Weight:	Proposed Resilience Score:
Thresholds:	Threshold-Based Score:	Your Score:
Less than 85%	1 (lowest resilience)	1 (lowest resilience)
85 to 90%	2	2
90 to 95%	3	3
Greater than 95%	4 (highest resilience)	4 (highest resilience)

#1157: Percentage of housing units with air conditioning

Definition: This indicator reflects the percentage of housing units with air conditioning.

Grouped with Indicators: #393, #728, #1170, #1387

Data Set(s):

Notes on Data Set(s):

Indicator Value:

Relevance:	Importance Weight:	Proposed Resilience Score:
Thresholds:	Threshold-Based Score:	Your Score:
Less than 70%	1 (lowest resilience)	1 (lowest resilience)
70 to 88%	2	2
88 to 94%	3	3
Greater than 94%	4 (highest resilience)	4 (highest resilience)

#1170: Percentage of population experiencing heat-related deaths

Definition: This indicator reflects the percentage of the population experiencing heat-related deaths.

Grouped with Indicators: #393, #728, #1157, #1387

Data Set(s):

Notes on Data Set(s):

Indicator Value:

Relevance:	*Importance Weight:*	*Proposed Resilience Score:*
Thresholds:	*Threshold-Based Score:*	*Your Score:*
Greater than 2.0%	1 (lowest resilience)	1 (lowest resilience)
1.0 to 2.0%	2	2
0.5 to 1.0%	3	3
Less than 0.5%	4 (highest resilience)	4 (highest resilience)

#1171: Percentage of population affected by food poisoning

Definition: This indicator reflects the percentage of population affected by food poisoning (i.e., *Salmonella* spp., unsafe drinking water).

Grouped with Indicators: #322, #676

Data Set(s):

Notes on Data Set(s):

Indicator Value:

Relevance:	Importance Weight:	Proposed Resilience Score:
Thresholds:	*Threshold-Based Score:*	*Your Score:*
Greater than 20%	1 (lowest resilience)	1 (lowest resilience)
15 to 20%	2	2
10 to 15%	3	3
Less than 10%	4 (highest resilience)	4 (highest resilience)

J.6. Telecommunications

The indicators below have been developed for the telecommunication sector. Indicators that are related are grouped together such that a single indicator from that group was considered a **primary indicator** and the remaining were considered **secondary indicators.** Primary indicators and nongrouped indicators are presented in the first half of this handout, followed by the secondary indicators.

Each indicator has a **definition**. Each question is flagged with one or more of the following gradual change climate stressor and/or extreme event climate stressor (from the urban resilience framework developed for this project):

Stressors

Gradual Changes	*Extreme Events*
Wind speed	Magnitude/duration of heat waves
Temperature	Drought intensity/duration
Precipitation	Flood magnitude/frequency
Sea level rise	Hurricane intensity/frequency
	Storm surge/flooding

Where it was possible to identify a data set that would provide data for the indicator for Washington, DC, **data sets** and associated **notes on available data** are included. Indicators are assigned a **proposed resilience score** on a scale of 1 (lowest resilience) to 4 (highest resilience).

For each indicator, please:

1. Discuss the **relevance** of the indicator to the telecommunication sector. (If unsure, please select the *not sure—remind me later* option.) Indicators may be selected as *yes (relevant)* on the basis of the stressors previously selected as being most relevant to Washington, DC or based on any other criteria. Secondary indicators may be considered if the primary indicator is not adequately defined or does not have available data set(s).
2. When possible, **data sets** for Washington, DC are provided where data were available. In some cases, no data sets were identified. Please suggest data sets that may be better than the data sets identified or where data gaps exist.
3. For indicators selected as *yes (relevant)*, discuss an **importance weight**, where 1 = not very important and 4 = very important.
4. Review the **proposed resilience score** (if provided), which is on a scale of 1 (lowest resilience) to 4 (highest resilience), for the indicator. If you disagree with this score, please discuss **your score** and indicate the reason for your disagreement.

PRIMARY INDICATORS AND NONGROUPED INDICATORS

#1433: Percentage of system capacity needed to carry baseline level of traffic

Definition: Percentage of system capacity needed to carry baseline level of traffic.

Grouped with Indicators: N/A

Data Set(s):

Notes on Data Set(s):

Indicator Value:

Relevance:	*Importance Weight:*	*Proposed Resilience Score:*
Thresholds:	*Threshold-Based Score:*	*Your Score:*
Greater than 70%	1 (lowest resilience)	1 (lowest resilience)
50 to 70%	2	2
30 to 50%	3	3
Less than 30%	4 (highest resilience)	4 (highest resilience)

#1434: Baseline percentage of water supply for telecommunication systems that comes from outside the metropolitan area

Definition:

Grouped with Indicators:

Data Set(s):

Notes on Data Set(s):

Indicator Value:

Relevance:	*Importance Weight:*	*Proposed Resilience Score:*
Thresholds:	*Threshold-Based Score:*	*Your Score:*
Greater than 50%	1 (lowest resilience)	1 (lowest resilience)
20 to 50%	2	2
5 to 20%	3	3
Less than 5%	4 (highest resilience)	4 (highest resilience)

#1435: Baseline percentage of energy supply for telecommunication systems that comes from outside the metropolitan area

Definition:

Grouped with Indicators:

Data Set(s):

Notes on Data Set(s):

Indicator Value:

Relevance:	*Importance Weight:*	*Proposed Resilience Score:*
Thresholds:	*Threshold-Based Score:*	*Your Score:*
Greater than 60%	1 (lowest resilience)	1 (lowest resilience)
30 to 60%	2	2
10 to 30%	3	3
Less than 10%	4 (highest resilience)	4 (highest resilience)

#1441: Percentage of community with access to FEMA emergency radio broadcasts

Definition: Percentage of community with access to FEMA emergency radio broadcasts.

Grouped with Indicators: N/A

Data Set(s):

Notes on Data Set(s):

Indicator Value:

Relevance: *Importance Weight:* *Proposed Resilience Score:*

Thresholds:	Threshold-Based Score:	Your Score:
Less than 80%	1 (lowest resilience)	1 (lowest resilience)
80 to 88%	2	2
88 to 96%	3	3
Greater than 96%	4 (highest resilience)	4 (highest resilience)

J.7. Transportation

The indicators below have been developed for the transportation sector. Indicators that are related are grouped together such that a single indicator from that group was considered a **primary indicator** and the remaining were considered **secondary indicators.** Primary indicators and nongrouped indicators are presented in the first half of this handout, followed by the secondary indicators.

Each indicator has a **definition**. Each question is flagged with one or more of the following gradual change climate stressor and/or extreme event climate stressor (from the urban resilience framework developed for this project):

Stressors

Gradual Changes	*Extreme Events*
Wind speed	Magnitude/duration of heat waves
Temperature	Drought intensity/duration
Precipitation	Flood magnitude/frequency
Sea level rise	Hurricane intensity/frequency
	Storm surge/flooding

Where it was possible to identify a data set that would provide data for the indicator for Washington, DC, **data sets** and associated **notes on available data** are included. Indicators are assigned a **proposed resilience score** on a scale of 1 (lowest resilience) to 4 (highest resilience).

For each indicator, please:

1. Discuss the **relevance** of the indicator to the transportation sector. (If unsure, please select the *not sure—remind me later* option.) Indicators may be selected as *yes (relevant)* on the basis of the stressors previously selected as being most relevant to Washington, DC or based on any other criteria. Secondary indicators may be considered if the primary indicator is not adequately defined or does not have available data set(s).
2. When possible, **data sets** for Washington, DC are provided where data were available. In some cases, no data sets were identified. Please suggest data sets that may be better than the data sets identified or where data gaps exist.
3. For indicators selected as *yes (relevant)*, discuss an **importance weight**, where 1 = not very important and 4 = very important.
4. Review the **proposed resilience score** (if provided), which is on a scale of 1 (lowest resilience) to 4 (highest resilience), for the indicator. If you disagree with this score, please discuss **your score** and indicate the reason for your disagreement.

PRIMARY INDICATORS AND NONGROUPED INDICATORS

#988: Walkability score

Definition: This indicator reflects the walkability score of the community (points out of 100).

Grouped with Indicators: #987, #1396, #1417

Data Set(s):

Notes on Data Set(s):

Indicator Value:

Relevance:	*Importance Weight:*	*Proposed Resilience Score:*
Thresholds:	*Threshold-Based Score:*	*Your Score:*
0 to 49 "car dependent"	1 (lowest resilience)	1 (lowest resilience)
50 to 69 "somewhat walkable"	2	2
70 to 89 "very walkable"	3	3
90 to 100 "walker's paradise"	4 (highest resilience)	4 (highest resilience)

#1402: Total annual hours of rail line closure due to heat and maintenance problems

Definition: This indicator measures (1) total annual hours that rail lines within the metropolitan transit system are closed due to heat kinks and (2) total annual hours that transit vehicles are unable to operate due to maintenance problems associated with extreme heat stress.

Grouped with Indicators: #1410

Data Set(s):

Notes on Data Set(s):

Indicator Value:

Relevance:	Importance Weight:	Proposed Resilience Score:
Thresholds:	Threshold-Based Score:	Your Score:
Greater than 6 hours	1 (lowest resilience)	1 (lowest resilience)
3 to 6 hours	2	2
1 to 3 hours	3	3
Less than 1 hour	4 (highest resilience)	4 (highest resilience)

#1404: Percentage of city culverts that are sized to meet future stormwater capacity requirements

Definition: This indicator measures the percentage of current culverts that cross transportation facilities in the metropolitan region that are sized to meet projected stormwater capacity requirements for 2030.

Grouped with Indicators: #1403

Data Set(s):

Notes on Data Set(s):

Indicator Value:

Relevance:	Importance Weight:	Proposed Resilience Score:
Thresholds:	Threshold-Based Score:	Your Score:
Less than 70%	1 (lowest resilience)	1 (lowest resilience)
70 to 85%	2	2
85 to 95%	3	3
Greater than 95%	4 (highest resilience)	4 (highest resilience)

#1412: Miles of pedestrian facilities per street mile

Definition: This indicator measures the miles of pedestrian facilities (sidewalks) per street mile.

Grouped with Indicators: #1413

Data Set(s):

Notes on Data Set(s):

Indicator Value:

Relevance:	Importance Weight:	Proposed Resilience Score:
Thresholds:	Threshold-Based Score:	Your Score:
Less than 0.5 miles of sidewalk to street miles	1 (lowest resilience)	1 (lowest resilience)
0.5 to 1.0 miles of sidewalk to street miles	2	2
1.0 to 2.0 miles of sidewalk to street miles	3	3
Greater than 2.0 miles of sidewalk to street miles	4 (highest resilience)	4 (highest resilience)

#1420: Intermodal passenger connectivity (percentage of terminals with at least one intermodal connection for the most common mode)

Definition: This indicator measures the percentage of active passenger terminals for the most common mode (e.g., rail, air, etc.) with at least one intermodal passenger connection. Intermodal connections allow passengers to use a combination of modes and give travelers additional transportation alternatives that unconnected, parallel systems do not offer.

Grouped with Indicators: #1419

Data Set(s):

Notes on Data Set(s):

Indicator Value:

Relevance:	*Importance Weight:*	*Proposed Resilience Score:*
Thresholds:	*Threshold-Based Score:*	*Your Score:*
Less than 55%	1 (lowest resilience)	1 (lowest resilience)
55 to 70%	2	2
70 to 85%	3	3
Greater than 85%	4 (highest resilience)	4 (highest resilience)

#985: Transport system user satisfaction

Definition: This indicator reflects the overall user satisfaction with the transport system. It is defined as the average user satisfaction with bus service, rail service, and the accuracy of passenger information displays.

Grouped with Indicators: N/A

Data Set(s):

Notes on Data Set(s):

Indicator Value:

Relevance:	Importance Weight:	Proposed Resilience Score:
Thresholds:	Threshold-Based Score:	Your Score:
0 to 20 (very or totally dissatisfied)	1 (lowest resilience)	1 (lowest resilience)
21 to 60 (somewhat dissastisfied)	2	2
61 to 80 (somewhat satisfied)	3	3
81 to 100 (very or totally satisfied)	4 (highest resilience)	4 (highest resilience)

#991: Percentage transport diversity

Definition: Highest public expenditure for a single mode of transportation as a percentage of the total expenditures for all transportation modes.

Grouped with Indicators: N/A

Relevance:	*Importance Weights:*
Yes (relevant)	1 (not very important)
No (not relevant)	2
Not sure—remind me later	3
	4 (very important)

Data Set(s):

Notes on Data Set(s):

Indicator Value:

Proposed Resilience Score:	*Your Score:*
	1 (lowest resilience)
	2
	3
	4 (highest resilience)

#1003: Mobility management (yearly congestion costs saved by operational treatments per capita)

Definition: Implementation of mobility management programs can address problems and increase transport system efficiency. This indicator reports on the yearly congestion costs saved by operational treatments (in billions of 2011 dollars). Operational treatments include freeway incident management, freeway ramp metering, arterial street signal coordination, arterial street access management, and high-occupancy vehicle lanes.

Grouped with Indicators: N/A

Data Set(s):

Notes on Data Set(s):

Indicator Value:

Relevance:	Importance Weight:	Proposed Resilience Score:
Thresholds:	Threshold-Based Score:	Your Score:
$2 to less than $10 per person	1 (lowest resilience)	1 (lowest resilience)
$10 to less than $18 per person	2	2
$18 to less than $32 per person	3	3
Greater than or equal to $32 per person	4 (highest resilience)	4 (highest resilience)

#1010: Community Livability Index

Definition: The Community Livability Index is the equally weighted average of the Community Service Indicator, the Crime Indicator, the Retail Opportunity Indicator, the Educational Indicator, the Environmental Quality Indicator, the Housing Affordability Indicator, and the Transit Livability Indicator. Details of the calculation are provided in Ripplinger et al. (2012; http://www.ugpti.org/pubs/pdf/DP262.pdf).

Grouped with Indicators: N/A

Data Set(s):

Notes on Data Set(s):

Indicator Value:

Relevance:	*Importance Weight:*	*Proposed Resilience Score:*
Thresholds:	*Threshold-Based Score:*	*Your Score:*
Less than 60 (most aspects of living are substantially constrained or severely restricted)	1 (lowest resilience)	1 (lowest resilience)
61 to 70 (negative factors have an impact on day-to-day living)	2	2
71 to 80 (day-to-day living is fine, in genera, but some aspects of life may entail problems)	3	3
81 to 100 (there are few, if any challenges to living standards)	4 (highest resilience)	4 (highest resilience)

#1399: Percentage of roads and railroads within the city that are located within 10 feet of water

Definition: This indicator measures the percentage of roadway miles and rail line miles that are within 10 feet of a body of water.

Grouped with Indicators: N/A

Relevance:	*Importance Weights:*
Yes (relevant)	1 (not very important)
No (not relevant)	2
Not sure—remind me later	3
	4 (very important)

Data Set(s):

Notes on Data Set(s):

Indicator Value:

Proposed Resilience Score:	*Your Score:*
	1 (lowest resilience)
	2
	3
	4 (highest resilience)

307

#1400: Percentage of roads and railroads within the city in the 500-year floodplain

Definition: This indicator measures the percentage of roadway miles and rail line miles that are within the 500-year floodplain.

Grouped with Indicators: N/A

Data Set(s):

Notes on Data Set(s):

Indicator Value:

Relevance:	*Importance Weight:*	*Proposed Resilience Score:*
Thresholds:	*Threshold-Based Score:*	*Your Score:*
Greater than 5%	1 (lowest resilience)	1 (lowest resilience)
2 to 5%	2	2
1 to 2%	3	3
Less than 1%	4 (highest resilience)	4 (highest resilience)

#1401: Percentage of roads and railroads within the city in the 100-year floodplain

Definition: This indicator measures the percentage of roadway miles and rail line miles that are within the 100-year floodplain.

Grouped with Indicators: N/A

Data Set(s):

Notes on Data Set(s):

Indicator Value:

Relevance:	Importance Weight:	Proposed Resilience Score:
Thresholds:	Threshold-Based Score:	Your Score:
Greater than 20%	1 (lowest resilience)	1 (lowest resilience)
10 to 20%	2	2
5 to 10%	3	3
Less than 5%	4 (highest resilience)	4 (highest resilience)

#1406: Percentage decline in repeat maintenance events

Definition: This indicator measures the percentage decline in repeat maintenance events, thereby representing a stable transportation system. The most recent transportation bill states that roadways and bridges subject to repeat maintenance must be studied so as to avoid repeated use of emergency funds for infrastructure that keeps getting damaged.

Grouped with Indicators: N/A

Data Set(s):

Notes on Data Set(s):

Indicator Value:

Relevance:	Importance Weight:	Proposed Resilience Score:
	1 (not very important)	
	2	
	3	
	4 (very important)	
Thresholds:	Threshold-Based Score:	Your Score:
Less than 10%	1 (lowest resilience)	1 (lowest resilience)
10 to 25%	2	2
25 to 50%	3	3
Greater than 50%	4 (highest resilience)	4 (highest resilience)

#1408: Percentage of bridges that are structurally deficient

Definition: This indicator measures the percentage of bridges that are structurally deficient. Bridges are considered structurally deficient if significant load-carrying elements are found to be in poor or worse condition due to deterioration or damage, or the adequacy of the waterway opening provided by the bridge is determined to be extremely insufficient to the point of causing intolerable traffic interruptions.

Grouped with Indicators: N/A

Data Set(s):

Notes on Data Set(s):

Indicator Value:

Relevance:	Importance Weight:	Proposed Resilience Score:
Thresholds:	Threshold-Based Score:	Your Score:
Greater than 10%	1 (lowest resilience)	1 (lowest resilience)
5 to 10%	2	2
2 to 5%	3	3
Less than 2%	4 (highest resilience)	4 (highest resilience)

#1411: Roadway connectivity (number of intersections per square mile)

Definition: This indicator measures the number of intersections per square mile.
Grouped with Indicators: N/A
Data Set(s):

Notes on Data Set(s):

Indicator Value:

Relevance:	Importance Weight:	Proposed Resilience Score:
Thresholds:	Threshold-Based Score:	Your Score:
Less than 80 intersections per square mile	1 (lowest resilience)	1 (lowest resilience)
80 to 250 intersections per square mile	2	2
250 to 290 intersections per square mile	3	3
Greater than 290 intersections per square mile	4 (highest resilience)	4 (highest resilience)

#1422: Average distance of all nonwork trips

Definition: This indicator measures the average distance from a given home to the nearest grocery store, high school, and health care facility (i.e., nonwork trips).

Grouped with Indicators: N/A

Data Set(s):

Notes on Data Set(s):

Indicator Value:

Relevance:	Importance Weight:	Proposed Resilience Score:
Thresholds:	Threshold-Based Score:	Your Score:
Less than 5 miles	1 (lowest resilience)	1 (lowest resilience)
5 to 10 miles	2	2
10 to 30 miles	3	3
Greater than 30 miles	4 (highest resilience)	4 (highest resilience)

#1426: City congestion rank

Definition: This indicator measures the congestion rank of the metropolitan area relative to all U.S. metropolitan areas.

Grouped with Indicators: N/A

Data Set(s):

Notes on Data Set(s):

Indicator Value:

Relevance:	Importance Weight:	Proposed Resilience Score:
Thresholds:	Threshold-Based Score:	Your Score:
1 to 25 (unitless rank)	1 (lowest resilience)	1 (lowest resilience)
26 to 50 (unitless rank)	2	2
51 to 75 (unitless rank)	3	3
76 to 100 (unitless rank)	4 (highest resilience)	4 (highest resilience)

#1429: Telework rank

Definition: This indicator measures the telework rank of the mtropolitan area relative to all other extralarge metropolitan areas in the U.S. The rank is based on the percentage of jobs within the metropolitan region that could be accomplished by telecommuting if employer policies were to permit it.

Grouped with Indicators: N/A

Data Set(s):

Notes on Data Set(s):

Indicator Value:

Relevance:	Importance Weight:	Proposed Resilience Score:
Thresholds:	Threshold-Based Score:	Your Score:
13 to 16 (unitless rank)	1 (lowest resilience)	1 (lowest resilience)
9 to 12 (unitless rank)	2	2
5 to 8 (unitless rank)	3	3
1 to 4 (unitless rank)	4 (highest resilience)	4 (highest resilience)

SECONDARY INDICATORS

#987: Employment accessibility (mean travel time to work relative to national average)

Definition: This indicator is defined as the mean travel time to work in a city relative to the U.S. average.

Grouped with Indicators: #988, #1396, #1417

Data Set(s):

Notes on Data Set(s):

Indicator Value:

Relevance:	Importance Weight:	Proposed Resilience Score:
Thresholds:	Threshold-Based Score:	Your Score:
Greater than 1.18 (unitless ratio)	1 (lowest resilience)	1 (lowest resilience)
0.98 to 1.18 (unitless ratio)	2	2
0.79 to less than 0.98 (unitless ratio)	3	3
Less than 0.79 (unitless ratio)	4 (highest resilience)	4 (highest resilience)

#1396: Percentage access to transportation stops

Definition: This indicator reflects the percentage of the population that is near a transit stop.
Grouped with Indicators: #987, 988, #1417
Data Set(s):

Notes on Data Set(s):

Indicator Value:

Relevance:	Importance Weight:	Proposed Resilience Score:
Thresholds:	Threshold-Based Score:	Your Score:
23 to 47%	1 (lowest resilience)	1 (lowest resilience)
48 to 63%	2	2
64 to 75%	3	3
76 to 100%	4 (highest resilience)	4 (highest resilience)

#1403: Percentage of city culverts that are sized to meet current stormwater capacity requirements

Definition: This indicator measures the percentage of current culverts that cross transportation facilities in the metropolitan region that are sized to meet current stormwater capacity requirements.

Grouped with Indicators: #1404

Data Set(s):

Notes on Data Set(s):

Indicator Value:

Relevance:	Importance Weight:	Proposed Resilience Score:
Yes (relevant)	1 (not very important)	
No (not relevant)	2	
	3	
	4 (very important)	
Thresholds:	*Threshold-Based Score:*	*Your Score:*
Less than 75%	1 (lowest resilience)	1 (lowest resilience)
75 to 90%	2	2
90 to 95%	3	3
Greater than 95%	4 (highest resilience)	4 (highest resilience)

#1410: Hours of passenger delay due to heat related issues

Definition: N/A

Grouped with Indicators: #1402

Relevance:	*Importance Weights:*
Yes (relevant)	1 (not very important)
No (not relevant)	2
Not sure—remind me later	3
	4 (very important)

Data Set(s):

Notes on Data Set(s):

Indicator Value:

Proposed Resilience Score:	*Your Score:*
	1 (lowest resilience)
	2
	3
	4 (highest resilience)

#1413: Percentage of short walkable sidewalks in urban areas

Definition: This indicator measures the percentage of sidewalks within the urban area that are less than 330 feet.

Grouped with Indicators: #1412

Data Set(s):

Notes on Data Set(s):

Indicator Value:

Relevance:	*Importance Weight:*	*Proposed Resilience Score:*
Thresholds:	*Threshold-Based Score:*	*Your Score:*
Less than 60%	1 (lowest resilience)	1 (lowest resilience)
60 to 75%	2	2
75 to 90%	3	3
Greater than 90%	4 (highest resilience)	4 (highest resilience)

#1417: Percentage funding spent on pedestrian/bicycle projects connected to community activity centers

Definition: Percentage of program funds spent on pedestrian or bicycle projects that include at least one connection to activity centers (e.g., schools; universities; downtown and employment districts; senior facilities; hospital/medical clinics; parks, recreation, and sporting; grocery stores; museums and tourist attractions).

Grouped with Indicators: #987, #988, #1396

Relevance:	*Importance Weights:*
Yes (relevant)	1 (not very important)
No (not relevant)	2
Not sure—remind me later	3
	4 (very important)

Data Set(s):

Notes on Data Set(s):

Indicator Value:

Proposed Resilience Score:	*Your Score:*
	1 (lowest resilience)
	2
	3
	4 (highest resilience)

#1419: Intermodal freight connectivity (ratio of intermodal connections used per year to individual modes)

Definition: This indicator measures the number of intermodal connections per year relative to distinct modes. Intermodal connections allow freight to use a combination of modes and give shippers additional transportation alternatives that unconnected, parallel systems do not offer.

Grouped with Indicators: #1420

Data Set(s):

Notes on Data Set(s):

Indicator Value:

Relevance:	Importance Weight:	Proposed Resilience Score:
Yes (relevant) No (not relevant)		
Thresholds:	*Threshold-Based Score:*	*Your Score:*
Less than 0.5 ratio of intermodal containers to individual carloads	1 (lowest resilience)	1 (lowest resilience)
0.5 to 1.0 ratio of intermodal containers to individual carloads	2	2
1 to 2 ratio of intermodal containers to individual carloads	3	3
Greater than 2 ratio of intermodal containers to individual carloads	4 (highest resilience)	4 (highest resilience)

J.8. Water

The indicators below have been developed for the water sector. Indicators that are related are grouped together such that a single indicator from that group was considered a **primary indicator** and the remaining were considered **secondary indicators**. Primary indicators and nongrouped indicators are presented in the first half of this handout, followed by the secondary indicators.

Each indicator has a **definition**. Each question is flagged with one or more of the following gradual change climate stressor and/or extreme event climate stressor (from the urban resilience framework developed for this project):

Stressors

Gradual Changes
- Wind speed
- Temperature
- Precipitation
- Sea level rise

Extreme Events
- Magnitude/duration of heat waves
- Drought intensity/duration
- Flood magnitude/frequency
- Hurricane intensity/frequency
- Storm surge/flooding

Where it was possible to identify a data set that would provide data for the indicator for Washington, DC, **data sets** and associated **notes on available data** are included. Indicators are assigned a **proposed resilience score** on a scale of 1 (lowest resilience) to 4 (highest resilience).

For each indicator, please:

1. Discuss the **relevance** of the indicator to the water sector. (If unsure, please select the *not sure—remind me later* option.) Indicators may be selected as *yes (relevant)* on the basis of the stressors previously selected as being most relevant to Washington, DC or based on any other criteria. Secondary indicators may be considered if the primary indicator is not adequately defined or does not have available data set(s).
2. When possible, **data sets** for Washington, DC are provided where data were available. In some cases, no data sets were identified. Please suggest data sets that may be better than the data sets identified or where data gaps exist.
3. For indicators selected as *yes (relevant)*, discuss an **importance weight**, where 1 = not very important and 4 = very important.
4. Review the **proposed resilience score** (if provided), which is on a scale of 1 (lowest resilience) to 4 (highest resilience), for the indicator. If you disagree with this score, please discuss **your score** and indicate the reason for your disagreement.

PRIMARY INDICATORS AND NONGROUPED INDICATORS

#1346: Percentage of infiltration and inflow (I/I) in wastewater

Definition: Water that enters the wastewater system through infiltration and inflow (I/I) as a percentage of total wastewater from all wastewater treatment plants in the city. Infiltration is the seepage of groundwater into sewer pipes through cracks, holes, joint failures, or faulty connections. Inflow is surface water that enters the wastewater system from yard, roof and footing drains, cross-connections with storm drains, downspouts, and through holes in manhole covers.

Grouped with Indicators: N/A

Data Set(s):

Notes on Data Set(s):

Indicator Value:

Relevance:	Importance Weight:	Proposed Resilience Score:
	1 (not very important)	
	2	
	3	
	4 (very important)	
Thresholds:	Threshold-Based Score:	Your Score:
Greater than 50%	1 (lowest resilience)	1 (lowest resilience)
35 to 50%	2	2
20 to 35%	3	3
Less than 20%	4 (highest resilience)	4 (highest resilience)

#1347: Wet weather flow bypass volume relative to the 5-year average

Definition: Volume of wastewater that bypassed treatment in an average year for all wastewater treatment plants divided by the 5-year average.

Grouped with Indicators: N/A

Data Set(s):

Notes on Data Set(s):

Indicator Value:

Relevance:	Importance Weight:	Proposed Resilience Score:
Thresholds:	Threshold-Based Score:	Your Score:
Greater than 2 (unitless ratio)	1 (lowest resilience)	1 (lowest resilience)
1 to 2 (unitless ratio)	2	2
1 (unitless ratio)	3	3
Less than 1 (unitless ratio)	4 (highest resilience)	4 (highest resilience)

#1428: Total number of Safe Drinking Water Act (SDWA) violations

Definition: This indicator measures the total number of SDWA violations over the last 5 years.

Grouped with Indicators: N/A

Data Set(s):

Notes on Data Set(s):

Indicator Value:

Relevance:	Importance Weight:	Proposed Resilience Score:
	1 (not very important)	
	2	
	3	
	4 (very important)	

Thresholds:	Threshold-Based Score:	Your Score:
Greater than 4 violations	1 (lowest resilience)	1 (lowest resilience)
3 to 4 violations	2	2
1 to 2 violations	3	3
0 violations	4 (highest resilience)	4 (highest resilience)

#1442: Ratio of water consumption to water availability

Definition: This indicator measures the fraction of available water that is currently consumed. It is calculated by dividing total water consumption by the total available water from surface water and groundwater sources.

Grouped with Indicators: N/A

Data Set(s):
1442

Notes on Data Set(s):

Indicator Value:

Relevance:	Importance Weight:	Proposed Resilience Score:
Thresholds:	Threshold-Based Score:	Your Score:
Greater than 0.20 (unitless ratio)	1 (lowest resilience)	1 (lowest resilience)
0.13 to 0.20 (unitless ratio)	2	2
0.06 to 0.13 (unitless ratio)	3	3
Less than 0.06 (unitless ratio)	4 (highest resilience)	4 (highest resilience)

#437: Percentage change in streamflow divided by percentage change in precipitation

Definition: This indicator reflects percentage change in streamflow (Q) divided by percentage change in precipitation (P) for 1,291 gauged watersheds across the continental U.S. from 1931 to 1988.

Grouped with Indicators: #1369

Data Set(s):

Notes on Data Set(s):

Indicator Value:

Relevance:	Importance Weight:	Proposed Resilience Score:
Yes (relevant)	1 (not very important)	
No (not relevant)	2	
	3	
	4 (very important)	
Thresholds:	Threshold-Based Score:	Your Score:
Greater than 3.0 (unitless ratio)	1 (lowest resilience)	1 (lowest resilience)
2.0 to 3.0 (unitless ratio)	2	2
1.0 to 2.0 (unitless ratio)	3	3
Less than 1.0 (unitless ratio)	4 (highest resilience)	4 (highest resilience)

#1369: Annual CV of unregulated streamflow

Definition: The coefficient of variation (CV) of unregulated streamflow is an indicator of annual streamflow variability. It is computed as the ratio of the standard deviation of unregulated annual streamflow (oQs) to the unregulated mean annual streamflow (QS)' (Hurd et al., 1999).

Grouped with Indicators: #437

Data Set(s):

Notes on Data Set(s):

Indicator Value:

Relevance:	Importance Weight:	Proposed Resilience Score:
Yes (relevant)	1 (not very important)	
No (not relevant)	2	
	3	
	4 (very important)	
Thresholds:	Threshold-Based Score:	Your Score:
Greater than 0.60 (unitless ratio)	1 (lowest resilience)	1 (lowest resilience)
0.40 to 0.60 (unitless ratio)	2	2
0.20 to 0.40 (unitless ratio)	3	3
Less than 0.20 (unitless ratio)	4 (highest resilience)	4 (highest resilience)

www.ingramcontent.com/pod-product-compliance
Lightning Source LLC
Chambersburg PA
CBHW080007210526
45170CB00015B/1866